Smart Urban Development

A Sreekumar
R. S. Mohan Kumar

2026
Agri Horti Press
New Delhi-110 084

ISBN: 978-93-49690-79-0 (Hardbound)

Published By

Agri Horti Press

New Delhi

E-mail: agrihortipress@gmail.com

Digitally Printed at : Replika Press Pvt. Ltd.

Preface

Smart Urban Development presents an in-depth exploration of the rapidly evolving concept of urbanization in the context of smart city initiatives, focusing on the integration of advanced technologies, sustainable practices, and efficient infrastructure management. In response to the challenges posed by rapid urban growth, environmental concerns, and the need for enhanced quality of life, this book examines how smart technologies are reshaping urban landscapes, making cities more resilient, sustainable, and livable.

This book begins with an introduction to the core principles of smart urban development, emphasizing the role of digital technologies in improving the efficiency of urban services such as transportation, energy, waste management, water supply, and healthcare. The text explores how the use of the Internet of Things (IoT), big data, artificial intelligence (AI), and other innovative solutions are driving the creation of smart cities that can better respond to the needs of their inhabitants while minimizing their environmental impact.

The early chapters provide a detailed understanding of the foundational technologies behind smart urban development. These include sensor networks, cloud computing, data analytics, and integrated communication systems, which allow for the seamless exchange of information across various urban sectors. The book highlights how these technologies enable real-time monitoring and management of urban resources, leading to more informed decision-making and efficient service delivery.

A key focus of the book is on the concept of sustainable urban development. As cities face growing pressures related to population density, resource scarcity, and climate change, the need for sustainability has become paramount. This book covers the strategies that urban planners and policymakers can adopt to create eco-friendly, energy-efficient, and resilient urban environments. Topics such as green building practices, renewable energy integration, sustainable transportation systems, and waste reduction techniques are discussed in detail. The goal is to offer readers a vision of how cities can achieve long-term sustainability without compromising the needs of future generations.

Urban mobility and transportation are explored as critical components of smart cities. The integration of smart transportation systems, including autonomous vehicles, electric mobility, and real-time traffic management, is analyzed in terms of their potential to reduce congestion, lower emissions, and improve overall mobility. Additionally, the book examines the challenges of ensuring equitable access to these advanced technologies for all urban residents, regardless of socioeconomic status.

Social inclusion and citizen engagement are central to the idea of smart urban development. This book highlights the importance of designing urban spaces that are inclusive, accessible, and responsive to the diverse needs of city dwellers. It also explores how digital platforms, open data, and participatory governance models can empower citizens to actively participate in the decision-making processes that shape their urban environments.

Moreover, Smart Urban Development addresses the challenges and opportunities that arise with the implementation of smart city initiatives. It considers issues such as data privacy, cybersecurity, urban inequality, and the digital divide. The book discusses the need for a balanced approach that integrates technological innovation with social, economic, and environmental considerations to ensure that the benefits of smart urban development are widely distributed.

The final chapters focus on case studies from cities around the world that have successfully implemented smart urban development initiatives. These examples provide practical insights into the real-world application of the concepts discussed in the book, illustrating the tangible benefits and challenges of creating smarter, more sustainable cities.

In conclusion, Smart Urban Development is designed to serve as an essential resource for urban planners, policymakers, students, and professionals interested in the future of urbanization. It provides a comprehensive framework for understanding how smart technologies can be leveraged to build cities that are not only technologically advanced but also environmentally sustainable, socially inclusive, and economically viable.

A Sreekumar

R. S. Mohan Kumar

Contents

Section 1
Framework for Smart Urban Development

Chapter-1

Towards a Generic Framework for Smart Cities

1. Introduction

The rising demand for living in cities is likely to accentuate sustainability chal- lenges, climate change and resource allocation. Cities constantly compete for inter- national investment to generate employment, revenue and funds for development, all leading to elevated energy consumption and CO_2 emissions. Cities also seek innovation and efficiency in reducing time, cost, and energy in delivering services: smart transportation, intelligent buildings, and smart infrastructure that would lead to low carbon city development. In fact, 80% of the world's gross domestic product is created in cities; urban citizens earn on average three times the income of their rural counterparts; and people living in larger cities tend to have smaller energy footprints and require fewer infrastructures, consume less resources, and have higher productivity levels. A city of 8 million has 15% more productivity and 15% less infrastructure needs than two cities of 4 million each [2].

There are several urbanization models that incorporate digital technologies to address some of the urbanization and sustainability challenges. While digital cities attempt to integrate digital technology into city's infrastructure, intelligent cities utilize digital city infrastructure to construct intelligent urban systems featuring intelligent buildings, transportation systems, hospitals, schools, public services.

By the same token, smart cities deploy intelligent urban systems to support socio- economic development and improve urban quality of life [3].

Smart city initiatives seek to overcome the limitations of traditional urban development that manages infrastructure systems in silos and leverage the pervasive character of data and services offered by digital technologies, such as cloud computing, the internet of things, open and big data. As such, different stakeholders, investors and citizens work to enhance existing services and provide new services. Smart city development is highly complex, challenging and context-specific.

Challenges arise from discourses of technologies and policies, failure to tackle urban sustainability challenges, and governance framework.

2. Smart city concept

Over the past two decades, the concept of "smart cities" has surfaced to address the economic and social life of first worldwide cities [4]. Put simply, a smart city is a community that uses different data gathering devices to disseminate information that is used to manage services efficiently such as traffic control, power plants, water supply networks, hospitals, and other community services [5]. Within this context, citizens are very important for city's development. To keep them engaged, real quality services have to be offered at reasonable cost.

Associated as it is with technology, the concept of "smart city" has superseded other versions: "information city", "digital city" and the "intelligent city". In fact, the "digital city" originates from an experiment in Amsterdam in 1994, with the aim of democratizing access to the internet. The "digital city" now refers to: a connected community that combines broadband communications infrastructure; flexible, service-oriented computing infrastructure based on open industry standards; and innovative services to meet the needs of governments and their employees, citizens and businesses [6].

Smart city has been widely studied and registered under ISO 37120 sustainable cities and communities. The indicators of smart city services and quality of life are set out in ISO 37122 and resilient city standards are prescribed in ISO 37123 (**Figure 1**).

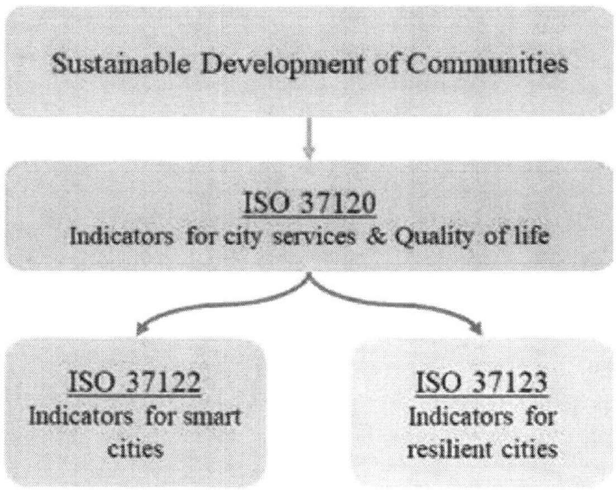

Figure 1. Smart city indicators and standards of sustainable development.

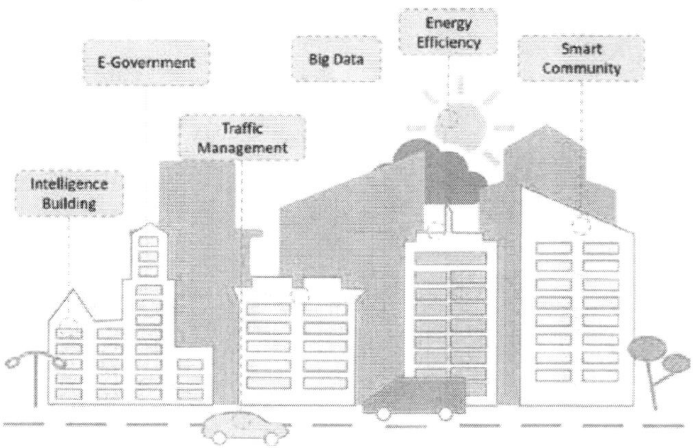

Figure 2. Basic components of smart city.

Indicators include, inter alia, economy, education, energy, climate change, finance, governance, health, housing, waste water and water quality. In the transportation sector for instance, data mining and sensing are used to obtain real-time data for managing duration of traffic light, traffic jam and accidents. It also potentially encourages mobility sharing through car, motorcycle and bicycle (**Figure 2**).

3. Smart city and carbon emissions

Because energy is central to smart city and low carbon cities, this section investigates the impact of urbanization on carbon emissions focusing on residential, commercial and industrial sectors, the major components of any city's land use.

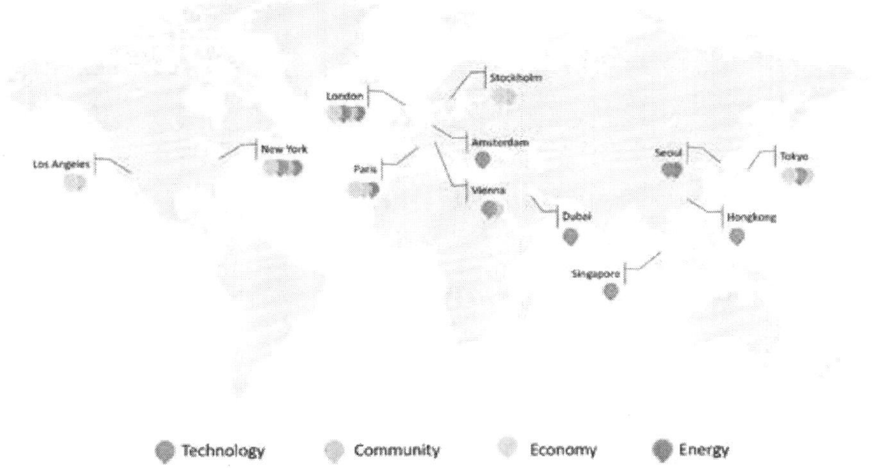

Figure 3. Significant sectors in selected global cities.

Azizalrahman and Hasyimi [7] have suggested a comparative analysis of low carbon cities in high income, upper-middle income and lower-middle income groups of countries. They have formulated an impact model of urban sector drivers on carbon emissions (USDM) to examine the relationship between urbanization, economic factors and carbon emissions and exposed urban dynamics of variables' interaction at city level. They found that most carbon emissions originating from the residential, commercial and public sectors are strongly influenced by energy consumption. Urbanization displays an inverse function with energy consumption and a positive correlation with economy. Based on IESE Cities in Motion Index 2018 [8], the performance of top global cities are measured and ranked based on dominant sectors which promote to sustainability (**Figure 3**).

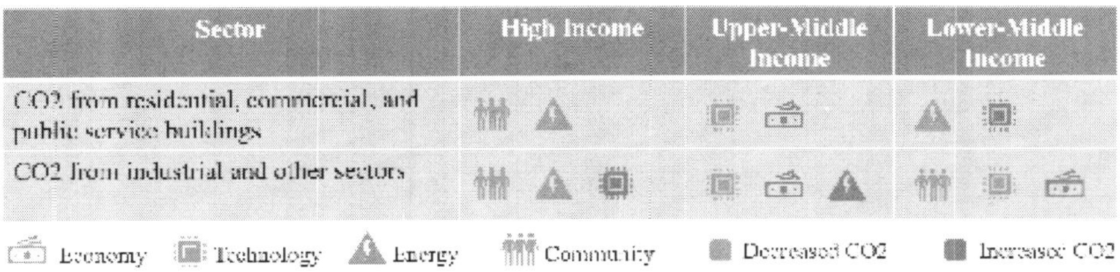

Figure 4. Effect of carbon emissions in smart cities of high, upper-middle, and lower-middle countries.

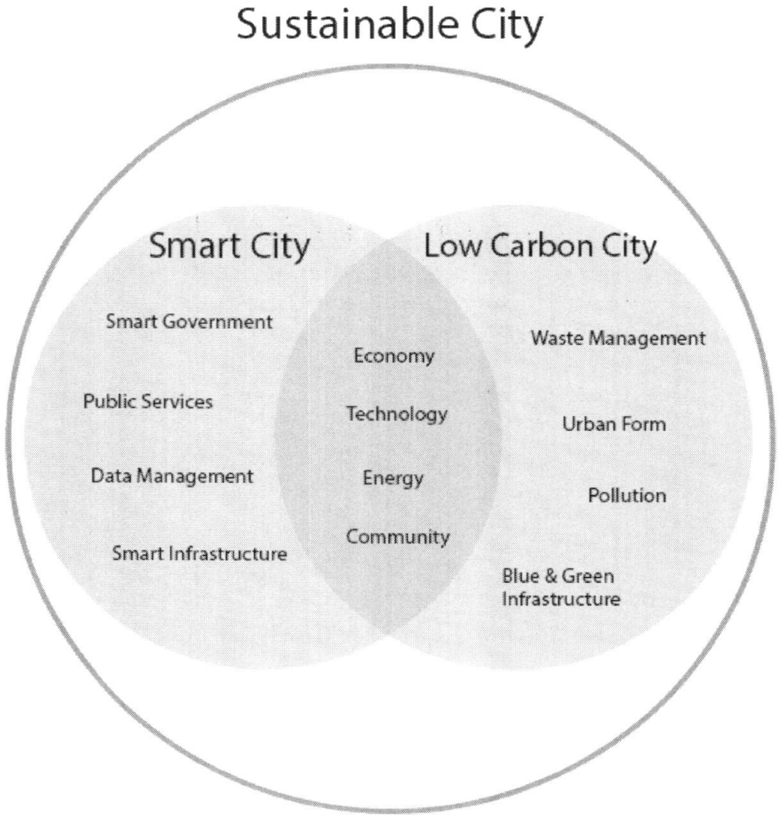

Figure 5. Commonalities between smart city and low carbon city in sustainability framework.

Based on their dominant characteristic, cities in lower middle-income countries have typical market towns that struggle on rapid urbanization. By contrast, cities in upper middle-income countries have typical production centres which focus on productivity. Cities in high income countries have become centres of finance and creative industries which face challenges of migrating firms to other regions (**Figure 4**).

Low carbon city and smart city are two forms of city development frameworks that purse sustainability. Low carbon city was established earlier than smart city in response to global warming and climate change. On the other hand, smart city has surfaced in the past decade to disseminate information and deploy technology solutions to improve efficiencies of city systems. Whereas low carbon city is mitigation purpose oriented, smart city is an adoption or adaptation targeted. Smart city has potentials to disseminate real data and record big data simultaneously thereby, enabling decision maker to track city system changes [9], see **Figure 5**.

Low carbon city framework has robust and clear targets, e.g., sulphur, nitrogen, and carbon emission levels. On the contrary, smart city has general; less specific targets that render measurement of smartness more difficult. Further, there is a widespread body of literature on low carbon city as opposed to relatively scant literature on smart cities. Some institutions have tried to develop evaluation models using sets of indicators to rank smart city performance such as smart cities ranking for Europe, world smart city government ranking, and the IESE Cities in Motion Index (CIMI) [8, 10–12].

4. Smart city framework

A smart city can be viewed within the wider perspective of sustainable city. The basic sectors include, amongst other things, technology, community, economy and energy which facilitate the development of a real concept of smart city. As such it gets closer to the definition of [11] who maintain that a city is smart when governance drives investment in human capital and IT infrastructure to achieve sustain- able development. The authors have constructed a fourfold framework for a typical smart city comprising technology, community, economy and energy to clearly distinguish between smart city and low carbon and sustainable cities (**Figure 6**).

a. Technology framework: ostensibly, smart cities are heavily dependent on the use of technology that is supported by technological infrastructure. These varied technologies are applied to diverse urban domains (e.g., economy, transportation, energy, environment, water management, waste disposal, education and healthcare, governance and public participation) to achieve efficiency and better management [9]. Within a Smart city context, information technology is not considered independently, but rather within wider physical and social systems that seek to deliver efficient service to people, business and government. It has become popular not only to smart cities, but also to engineering firms seeking innovation and investment opportunities for physical urban and infrastructure development.

b. Community framework: communities are central to city's intelligence as exemplified by human activities, innovation and knowledge. Human and social capital drives city's economy and technology deployment. Their power lies in effective creation of economic, cultural, social environment and formation of public opinion. Through participatory function, communities can influence policy formulation and decision making, such as redistribution of public finance and increasing the transparency of public expenditure. Representatives of cities, policy and decision makers should aim to reach consensus with the community on smart urban development [13].

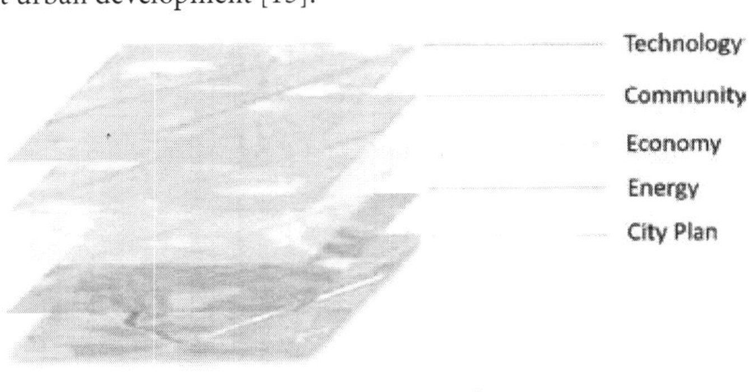

Figure 6. Basic smart city's sectors.

c. Economy framework: knowledge and digital economy are essential drivers of the smart city discourse. The terms "knowledge-based economy" refer to an economy where more knowledge-intensive than labour-intensive activities take place. It played a significant role in the emergence of the idea of smart cities; it is one of the two strands of thinking that formed the current ideas about what a smart city is, how it works, and what it can do. Moreover, smart city changes people's behaviour in purchasing from traditional to online transaction. It increases e-money usage, encourages store owners to react to this condition with some changes in their business models, etc.

d. Energy framework: smart cities seek to develop smart energy infrastructure, disseminate data to create efficiencies, leverage economic development, and enhance quality of life. A smart city features, inter alia, smart street lighting, intelligent buildings, smart mobility and power grid. The common thread is energy, economics and impact on cities. However, smart cities seem to have shifted attention away from environmental problems, climate change and carbon emissions to infrastructure and information usage and sharing.

4.1. Proposed smart city framework

A generic framework for smart cities is proposed comprising: (1) goal, (2) conceptualization, (3) assessment, and (4) implication. This model is useful to address smart city transformation that leads to sustainability. It affords a summary of complex transformation processes that are needed for cities seeking to be smart (**Figure 7**).

4.2. Smart city criteria and indicators

Common performance measurement methods use scoring methods which assess the current city condition. Here, the authors have used quantitative indicators used in the proposed model to create a generic framework to increase objectivity and realism. The indicators were obtained from several sources: ISO 337122, smart city in Europe, and generic model for low-carbon city [10]. The authors have initiated gathering of data for the basic sectors of the smart city: technology, community, economy and energy for which 20 key performance indicators (KPIs) were selected. For modelling purpose, the KPIs were then categorized under six urban development sectors: competitiveness, energy, mobility, urban management, urban living and waste management. The selected indicators can be seen in **Figure 8**.

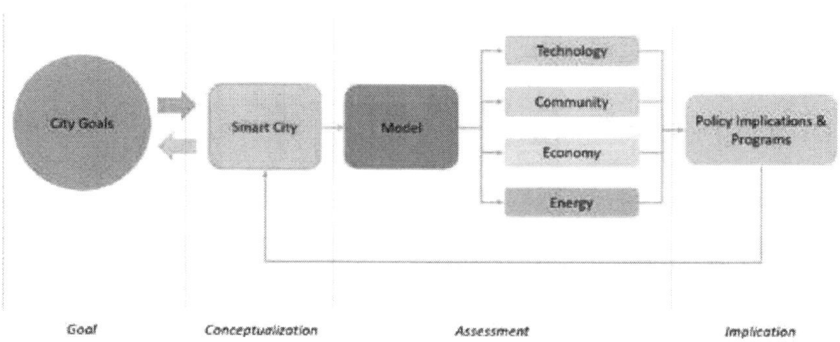

Figure 7. Proposed smart city framework.

Quantifiable indicators under each criterion are then selected to measure smart city performance and compare it with the benchmarks [14]. Benchmark setting is important because it aims to sufficiently differentiate between cities of various performance. Benchmarks were derived from multiple sources: (1) World Bank and WHO; (2) top city performances, such as green city index; (3) International targets for developed countries set out by EU (Table 1).

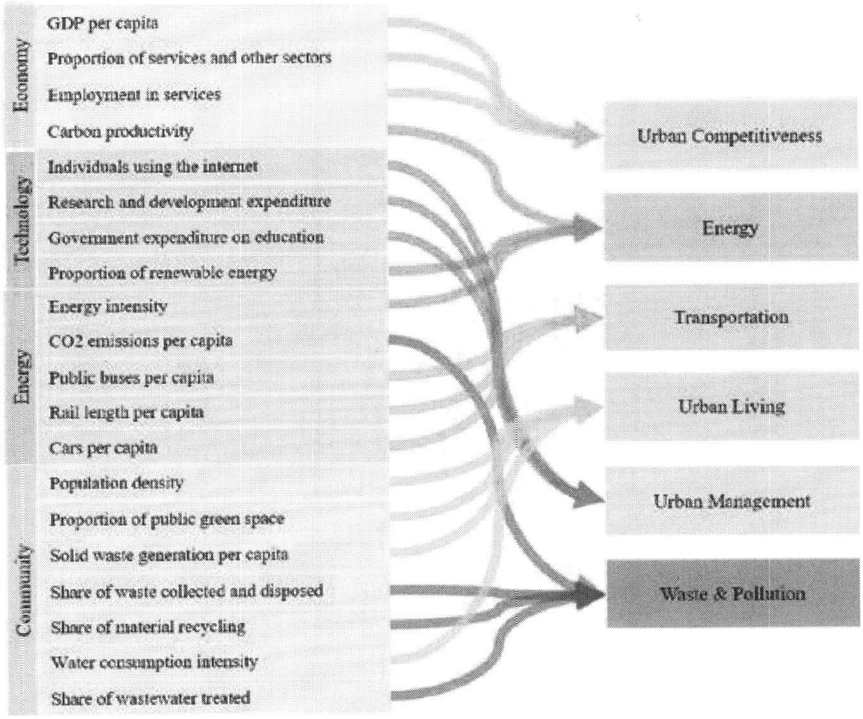

Figure 8. Smart city indicators and categorization.

Table 1. Smart city indicators and benchmarks.

Category	Indicator	Effect	Unit of measurement	Benchmark value	Source
Competitiveness	GDP per capita	+	$/capita	25,616	[15]
	Economy: services and other activity	+	% of gross value added	60	[16]
	Employment in services	+	% employed	60	[16]
Energy	Carbon productivity	+	USD/ton	8244	[17]
	Proportion of renewable energy	+	%	10	[17]
	Energy intensity	–	MJ/USD	4	[17]

Transportation	Public buses per capita	+	buses/million persons	694	[17]
	Rail length per capita	+	km/million persons	40	[17]
	Cars per capita	–	Private cars / persons	0.39	[17]
Urban Living	Proportion of public green space	+	%	35	[18]
	Population density	+	People/km^2	4236.1	[19]
	Solid waste generation per capita	–	Kg/capita/day	0.8	[20]
	Water consumption intensity	–	L/capita/day	102	[21]
Management	Education: government expenditure	+	% of GDP	3	[16]
	Individuals using the internet	+	per 100 inhabitants	70	[16]
	Research and development expenditure	+	% of GDP	1.5	[16]
Waste and pollution	CO_2 emission per capita	+	Ton/person	2.19	[21]
	Share of waste collected and adequately disposed	+	%	80	[20]
	Share of material recycling	+	%	30	[22]
	Share of wastewater treated	+	%	75	[21]

A multi-criteria evaluation model has been proposed by modifying the frame- work of Azizalrahman and Hasyimi [23].

The equation of data normalization is set out in Eqs. (1) and (2).

$$y_i = \frac{x_i - x_b}{} \quad (1)$$

$$y_i = \frac{x_b - x_i}{} \quad (2)$$

where y_i is normalized data of assessed object on i indicator, x_i is original value of the object on ith indicator, x_b is benchmark value of ith indicator. While Eq. (1) is used for indicators with positive effects, Eq. (2) is used for indicators with negative effects. This calculation will produce the score from minimum (−1) to maximum 1 (**Table 2**).

For better performance presentation, the standardization by score conversion to 0–100 could be seen in Eq. (3).

$$S_c = 50\ (y_i + 1) \quad (3)$$

Where S_c denotes the average score per category. S_T defines the average total score as shown in Eq. (4).

$$S_T = \left(\sum_{c=1}^{n} S_c\right)/6 \tag{4}$$

To obtain an average score S_T, an equal weight is assigned to 6 categories, the result of which features a smart city scale 0–100, from: unsustainable (0–9); high carbon (10–29); neutral (30–49); low carbon (50–69); smart (70–89) and sustain- able (90–100) as illustrated in **Figure 9**.

Table 2. Proposed multi-criteria evaluation model for smart city.

Indicator	Unit of measurement	Formula	y_{ic}	S_c
GDP Per capita	$/capita	$y_i = \frac{...-25616}{25616}$		
Economy: services and other activity	% of gross value added	$y_i = \frac{...-60}{60}$		
Employment in services	% employed	$y_i = \frac{...-60}{60}$		
Carbon productivity	USD/ton	$y_i = \frac{...-8244}{8244}$		
Proportion of renewable energy	%	$y_i = \frac{...-10}{10}$		
Energy intensity	MJ/USD	$y_i = \frac{4-...}{4}$		
Public buses per capita	buses/million persons	$y_i = \frac{...-694}{694}$		
Rail length per capita	km/million persons	$y_i = \frac{...40}{40}$		
Cars per capita	Private cars/persons	$y_i = \frac{0.39-...}{0.39}$		
Proportion of public green space	%	$y_i = \frac{...-35}{35}$		...
Population density	People/km^2	$y_i = \frac{...-4236.1}{4236.1}$		
Solid waste generation per capita	Kg/capita/day	$y_i = \frac{0.8-...}{0.8}$		
Water consumption intensity	L/capita/day	$y_i = \frac{102-...}{102}$		
Education: government expenditure	% of GDP	$y_i = \frac{...-3}{3}$		
Individuals using the internet	per 100 inhabitants	$y_i = \frac{...-70}{70}$		
Research and development expenditure	% of GDP	$y_i = \frac{...-1.5}{1.5}$		
CO_2 emission per capita	Ton/person	$y_i = \frac{2.19-...}{2.19}$		
Share of waste collected and adequately disposed	%	$y_i = \frac{...-80}{80}$		
Share of material recycling	%	$y_i = \frac{...-30}{30}$		
Share of wastewater treated	%	$y_i = \frac{...-75}{75}$		
Average (S_T)				...

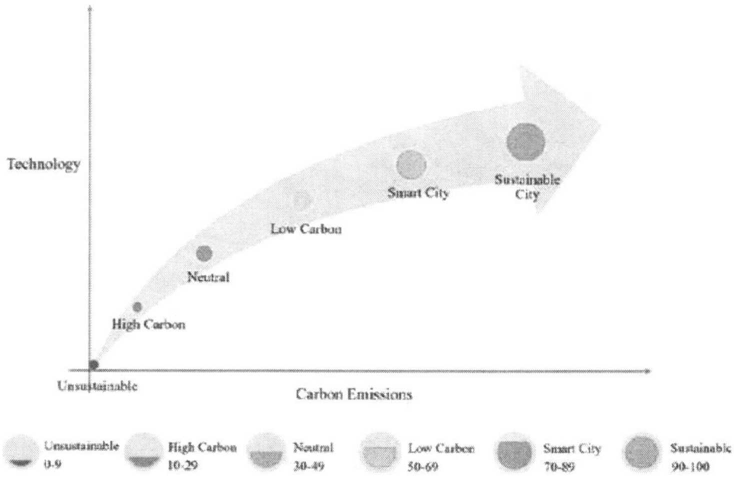

Figure 9. Smart city pathway to sustainability.

5. Smart city model and testing

The proposed model is tested on four cities: Vienna, London, New York, and Tokyo, the result of which can be seen in **Table 3**. The pilot cities are selected based on the good performance in technology sector based on IESE Cities in Motion Index 2018.

From the figure above, we can see that from four pilot cities, Vienna, London and New York are categorized as smart city. On the other side, Tokyo is low carbon city. The above scores were transformed into smart city metrics (**Figure 10**).

Smart city metrics help us summarize a detailed analysis for city's performance by sector. Through this presentation, the strength and weakness of each sector can be easily identified and promoted to achieve the desired targets. Vienna, a global tourism destination, has a very good performance in transportation and city management. Vienna has become a city of high mobility systems such as smart buses, smart ride, smart sharing, smart public transport, and eMorail to mention but a few. Moreover, Vienna has a peaceful balance between the city and green areas which account for half of the city's total area [24]. Therefore, the city is a leading smart city.

London and New York are examples of global cities with multiple central functions and populous agglomerations. Both have a strong performance in urban competitiveness and management. As centres of global trade and economy, London and New York have focused on, amongst other things, technology, human resource development, quality of urban living, and waste management.

London proved how smart the city could be by establishing London Datastore and innovation in transportation known as Heathrow pods; building up intelligent road network; facilitating trade with digital money; and making use of new tech- nology in reusing waste heat from underground chambers and sub-ways. London also executed the innovative program named as "Innovate18" which attempted to rejuvenate the old railway network [25].

By the same token, New York attempted to be a smart city by canvassing the concept of equitable city—a city where anyone and everyone has access to facilities justly. Being the economic hub of the world, the city is continuously engaged in delivering smart innovations. Current initiatives include reduction of greenhouse gases, fair management of water and energy, smart protection of public health increasing mortality rate and tech-based plans to make the city safer. Further, New York aims to set up strategies and policies to successfully actualise the connected devices and internet of things (IoT) [26].

Table 3. Result of smart city model on the pilot cities.

	Vienna	London	New York	Tokyo
Competitiveness	68	78	78	66
Energy	83	68	62	68
Transportation	83	84	79	75
Urban living	58	60	67	63
Management	62	66	60	63
Waste	69	73	72	70
Average score	71	71	70	68
Category	Smart city	Smart city	Smart city	Low carbon

Figure 10. Smart city metrics.

Tokyo on the other hand, is categorized as a low carbon city and is being trans- formed to a smart city. In the last few years, Tokyo has unveiled a chain of environment friendly initiatives which include: solid waste reduction through technology, encouragement of large-scale recycling plants and rain water harvesting, rooftop planting of trees and herbs which helps in absorbing carbon dioxide, adoption of energy efficient photovoltaic solar panels, and launch of Tokyo Super Eco Town [27].

References

1. Basiri M, Azim AZ, Farrokhi M. Smart city solution for sustainable urban development. European Journal of Sustainable Development. 2017;6(1):71-84
2. Urban Land Institute. Supporting Smart Urban Development: Successful Investing in Density. London; 2018
3. Roll G, Carriero D. United smart cities: Smart urban solutions for transition and developing countries. Vienna; 2015
4. McLaren D, Agyeman J. Sharing Cities: A Case for Truly Smart and Sustainable Cities. Cambridge: MIT Press; 2015
5. Musa S. Smart City Roadmap. San Francisco: Academia. Edu.; 2016
6. Giffinger R, Fertner C, Kramar H, Kalasek R, Pichler-Milanović N, Meijers E. Smart cities Ranking of European medium-sized cities. Vienna; 2007
7. Azizalrahman H, Hasyimi V. A model for urban sector drivers of carbon emissions. Sustainable Cities and Society 2019. October 2018;44:46-55
8. Berrone P, Ricart JE, Carrasco C, Figueras AID. IESE Cities in Motion Index 2018. Barcelona; 2018
9. Elias S, Krogstie J. Smart sustainable cities of the future: An extensive interdisciplinary literature review. Sustainable Cities and Society. 2017;31:183-212
10. Giffinger R, Gudrun H. Smart cities ranking: An effective instrument for the positioning of the cities. Architecture City, Environment 2010. February 2010;12:7-25
11. Caragliu A, Del Bo C, Nijkamp P. Smart Cities in Europe. Journal of Urban Technology. 2011;18(2):65-82
12. Eden Strategy Institute. Smart city governments. Singapore; 2018:12-18
13. Borsekova K, Vanova A, Vitalisova K. The power of communities in smart urban development. Procedia-Social and Behavioral Sciences. 2016;223:51-57
14. Tan S, Yang J, Yan J. Development of the low-Carbon city indicator (LCCI) framework. Energy Procedia. 2015;75(Lcci):2516-2522
15. World Bank. New Country Classifications [Online]. 2013. Available from: www. worldbank.org/news/new- country-classifications [Accessed: 08 Augest 2015]
16. United Nations. United Nations Data [Online]. 2018. Available from: www. data.un.org [Accessed: 02 March 2019]

17. Siemens. The green city index a summary of the green city index research series [Online]. 2015. Available from: http://www.siemens.com/press/pool/de/ events/2012/ corporate/2012-06-rio20/gci- report-e.pdf [Accessed: 13 Augest 2015]
18. Wien International. Environmental city Vienna—50% green space [Online]. 2015. Available from: https://www. wieninternational.at/en/content/ environmental-city-vienna-50-green- space-en [Accessed: 16 Aug 2015]
19. MERCER. Quality of living rangkings [Online]. 2015. Available from: https://www. imercer.com/ content/quality-of-living.aspx [Accessed: 11 Augest 2015]
20. World Bank. What a Waste a Global Review of Solid Waste Management. Washington: 2012 Zhou N, He G, Williams C, Fridley
21. D. ELITE cities: A low-carbon eco-city evaluation tool for China. Ecological Indicators. 2015;48:448-456
22. F. E. and E. M. A. (ADEME). Benchmarking study of waste treatment performance indicators. Final report— summary [Online]. 2013. Available from: http://www. franceintheus.org/ greenfrance/spip.php?article5126 [Accessed: 12 Augest 2015]
23. Azizalrahman H, Hasyimi V. Towards a generic multi-criteria evaluation model for low carbon cities. Sustainable Cities and Society. 2018;**39**:275-282
24. smartcity. Vienna—The Smartest City On The Planet [Online]. 2017. Available from: https://www.smartcity. press/viennas-smart-city-initiatives/ [Accessed: 09 March 2019]
25. smartcity. London—The Dawn Of Tech-rich Life Is Here [Online]. 2017. Available from: https://www.smartcity. press/londons-smart-city-initiatives/ [Accessed: 09 March 2019]
26. smartcity. The Equitable City—A New Name For New York [Online]. 2017. Available from: https://www.smartcity. press/new-yorks-smart-city-initiatives/ [Accessed: 09 March 2019]
27. smartcity. What Makes Tokyo The Greenest City Of Asia-Pacific Region [Online]. 2017. Available from: https:// www.smartcity.press/asia-pacifics- greenest-city-tokyo/ [Accessed: 09 March 2019]

Chapter-2

The Impact of Institutional and Political Factors on Timely Adoption of Local Community Budgets

1. Introduction

The Constitution of the Republic of Slovenia in Article 138 states that residents of Slovenia exercise local self-government in municipalities and other local communities. Article 139 further specifies that municipalities are self-governing local communities. The territory of a municipality comprises a settlement or several settlements bound together by the common needs and interests of the residents. A municipality is established by law following a referendum, by which the will of the residents in a given territory is determined. The territory of the municipality is also defined by the law. According to Article 140 of the Constitution, which defines the scope of local self-government, competencies of a municipality comprise local affairs which may be regulated by the municipality autonomously and which affect only the residents of the municipality. By law, the state may transfer to municipalities the performance of specific duties within the state competence, if it also provides the financial resources to enable such. "In recent years, both Russian and foreign researchers have given much attention to the development of local government" [1].

Local self-government is one of the fundamental principles in the Constitution of the Republic of Slovenia, with Article 9 stating: "Local self-government in Slovenia is guaranteed" [2].

The financing of the performance of duties within the municipality competence is regulated by the Financing of Municipalities Act, which in Article 3 states that the municipality financing shall be based on the principles of local self-government, mainly on the principle of proportionality of sources of financing with the municipality tasks and on the principle of independence of municipalities in financing municipality tasks. In Article 5, the act separates the financing of tasks within the state competence from the financing of tasks that municipalities must perform as part of their competencies, defined by the relevant laws. When drawing up the local community budget, the municipalities must comply with the applicable legal framework on public finance and accounting.

For a local community to be able to perform all the tasks in a certain calendar year, it must have the funds for its own operations and be financially independent. This means that it must adopt a budget, i.e., a financial plan it would follow in that year in time. Municipalities may also borrow money for individual tasks and investments, provided they comply with the legal framework.

Budget is a yearly financial plan, adopted in advance, in which the state or local community plans its revenue and expenses for fulfilling the needs of the public, with the municipal budgets being adopted as ordinances and the state budget as a legal act sui generis [3].

Drawing up a budget is an extremely important task of central and local authorities' executive bodies. This topic is also stressed in the works of George and John [4] and Copus [5]. Initiating and drawing up the budget in individual countries are generally regulated by budget laws, which specify the rights and obligations of individual budgetary authorities [6]. Under the law, the mayor represents the municipality and proposes the municipality budget to the municipal council for adoption, so it is important for the mayor to approach budgeting with all responsibility. McIlquham- Schmidt [7] identifies strategic planning as an effective way for improving the performance. However, a closer empirical review of the literature on the topic shows a greater diversity of findings. Some studies show a positive correlation between strategic planning and performance, which includes the relationship between employees [8], while others have found no correlation between the two [9]. These authors define planning differently, with the main point being the planning of future operations in companies as well as local communities, by focusing on 1- or 2-year plans for achieving the set goals, whose content and financial value are specified in the budget.

According to Shack [10], surveys have shown that when drawing up the budget at the state level, the level of political involvement increases. Municipal council members, on the other hand, are usually less intent on getting involved in local community budget planning.

A study from Denmark has shown that while the councilors listen to the citizens' comments and suggestions, they do not necessarily forward and defend them in the further process. Many local politicians are very interested in having a dialog with citizens, as this increases their chances of getting re-elected [11].

Budget is the result of a compromise between the interests of political parties and other stakeholders [3]. The weight of individual financial sources (own sources, state funding, borrowing) in a local community affects the adoption of the budget.

Institutional and political factors play a big role in the timely adoption of the local community's budget in Slovenia.

"Politicians are generally less interested in establishing performance municipal- ity budgets than top managers, particularly chief financial officers" [12]. So, we claim that collaboration between the mayor and finance director in drawing up the budget is needed, since finance directors play the main role in preparing the municipal budget. We take collaboration between the mayor and finance director as an example of an institutional factor. Clarity in informing the municipal council is an example of a political factor, with the council as the local community's decision- making body and the mayor all being elected. Findings in other research have shown that "political strength have a significant effect on the level of transparency in management of public resources" [13]. Other findings stated that "most projects are developed in narrow political circles including mayors, councils, municipal CEOs and semi-public organizations" [14].

A municipal budget is adopted for a financial year that starts and ends on the same dates as the state budget financial year. The mayor must present a draft budget to the municipal council within 30 days of the state budget being presented to the National Assembly. In the year when regular local elections are held, the mayor has 60 days to present the draft budget after the council is elected. A municipal budget is adopted for a financial year that starts and ends on the same dates as the state budget financial year. If the budget is not adopted before the start of the relevant year, it is deemed as not adopted on time. In this case, the mayor makes a decision on temporary financing for a period of up to 3 months, which can be extended by the council adopting a decision at the mayor's proposal.

Finance directors are present at council sessions when budgets are being adopted and have insight in the clarity of information exchanged between the mayor and the municipal council. The clarity of information provided to the municipal council affects the timely adoption of the budget. Out of 24 Council of Europe countries, 8 have adopted laws regarding collaboration between the mayor and politics (represented by the municipal council) in adopting the budget, specifying the processes and deadlines for preparing drafts. In general, they stipulate that the mayor must inform the representatives of people (the municipal council selected through the public vote) about the budget's basic premises and assumptions [11]. Mouritzen and Svara [15] closely studied how the mayor and the politics affect each other, as well as how political and executive actors impacted the local legislative body (Haček, Kukovič, and Grabner) [16].

2. The purpose and methods of work

This paper aims to examine the impact that institutional and political factors have on the timely adoption of a local community's budget, with a special focus on the adoption of budgets in 2016 in Slovenia. We will use the theoretical framework and similar research outcomes to compile recommendations for local communities on how to adopt budgets on time, which then in turn allows the local communities to finance their work and perform their tasks. If the local community's budget is not adopted on time, only temporary financing is provided to the community.

Our main thesis is that local communities face institutional and political factors when adopting the local community's budget, which affect the timely adoption of the budget and subsequently the performance of the tasks that are carried out by the municipality, i.e., the local community. The main thesis is based on several years of experience and understanding of the issues related to adopting local community budgets. Other similar authors' empirical research has already pointed to the issues related to local budgets. As we did not find comparable research in the literature to ours, we fill the research gap both empirically and theoretically.

We will check the selected factors, which we believe affect the budget adoption, based on the data collected in our survey sent out to finance directors in 212 municipalities, i.e., local communities in Slovenia. Different authors have already empirically proven that some of them affect the adoption of the budget. We will test the following hypothesis:

Hypothesis 1: Trust, openness, and daily collaboration between the mayor and finance director (institutional factor) are important factors that affect the timely adoption of the budget.

Hypothesis 2: The clarity of information the mayor presents to the council (political factor) is an important factor in the timely adoption of the local community's budget.

3. Survey results and discussion: trust, openness, and daily collaboration between the mayor and finance director

We used binary logistic regression to test hypothesis 1 that trust, openness, and daily collaboration between the mayor and finance director are important factors that affect the timely adoption of the budget. In our survey, we asked finance directors in municipalities to assess the elements of their collaboration with the mayor from the aspects of trust, openness, and daily collaboration and their impact on the adoption of the local community's budget. They rated the elements on a 5-point Likert scale, with 1 meaning inadequate and 5 excellent.

We first analyzed their responses on trust, openness, and daily collaboration between the mayor and finance director using the Mann–Whitney U test. We compared the ratings for trust, openness, and daily collaboration between the mayor and finance director in local communities where the budget was adopted on time and in those where it was not.

We further assessed the impact that the quality of communication between the finance director and mayor had on the timely adoption of the budget using binary logistic regression.

Based on Mann–Whitney U test (**Table 1**), we have concluded that all three examined elements have a statistically significant effect on the timely adoption of the budget. With the risk below 0.1%, we can confirm that all three elements linked to the quality of communication have a statistically significant effect, with the score for trust, openness, and daily collaboration higher in municipalities that adopted their budgets on time.

Daily collaboration between the mayor and finance director has a statistically significant effect on the timely adoption of the budget (U = 866.000; p = 0.000). As a result, we accept hypothesis 1 that trust, openness, and daily collaboration between the mayor and finance director are important factors that affect the timely adoption of the budget.

In municipalities where the budget was not adopted on time, the score for trust, openness, and daily collaboration between the mayor and finance director was lower. The average score for trust between the mayor and finance director in local communities where the budget was not adopted on time was 3.41 ± 0.923 (Me = 4). The average score for trust between the mayor and finance director in local communities where the budget was adopted on time was 4.37 ± 0.893 (Me = 5).

The average score for openness between the mayor and finance director in local communities where the budget was not adopted on time was 3.13 ± 0.795 (Me = 3).

Table 1. The impact of trust, openness, and daily collaboration between the mayor and finance director on the timely adoption of the budget.

Was your 2014 budget adopted before 1 January 2014?		N	Average range	Sum of ranges	Mann- Whitney U test	p
Trust	Yes	78	74.09	5779	734.00	<0.001
	No	44	39.18	1724		
	Total	122				
Openness	Yes	78	74.74	5830	683.00	<0.001

	No	44	38.02	1673		
	Total	122				
Daily collaboration	Yes	78	72.4	5647	866.00	<0.001
	No	44	42.18	1856		
	Total	122				

Table 2. Correlation between the scores for trust, openness, and daily collaboration.

		Trust	Openness	Daily collaboration
Trust	Correlation coefficient	1.000	0.818	0.617**
	p	.	0.000	0.000
	N	122	122	122
Openness	Correlation coefficient	0.818**	1.000	0.722**
	p	0.000	.	0.000
	N	122	122	122
Daily collaboration	Correlation coefficient	0.617**	0.722**	1.000
	p	0.000	0.000	.
	N	122	122	122

***The correlation is statistically significant at the 0.01 level (2-reps).*

The average score for openness between the mayor and finance director in local communities where the budget was adopted on time was 4.18 ± 0.977 (Me = 4). The average score for daily collaboration between the mayor and finance director in local communities where the budget was not adopted on time was 2.77 ± 1.008 (Me = 3). The average score for daily collaboration between the mayor and finance director in local communities where the budget was adopted on time was 3.78 ± 1.158 (Me = 4).

We first wanted to continue by conducting a binary logistic regression to assess the effect that trust, openness, and daily collaboration between the mayor and finance director have on the probability that the budget would be adopted on time. When checking for multicollinearity, we observed a moderate to strong correlation between all three input variables, which should be mutually independent (**Table 2**). Multicollinearity in a binary logistic regression could result in biased assessments and overblown standard errors. We have tested the appropriateness of the constructed binary logistic regression model, which included all three scores for communication as independent variables, with the Hosmer-Lemeshow test (**Table 3**), which revealed a statistically significant difference between the number of planned and announced budgets adopted on time ($\chi^2(7) = 17.595$, $p < 0.014$). We can therefore deduct that the constructed model was not appropriate.

Table 3. Results of binary logistic regression forecasting the timeliness of adopting the budget based on trust, openness, and collaboration.

Variable		**Score B**	**Standard error assessment**	**p**	**Odds ratio Exp(B)**	**95% confidence interval for Exp(B)**	
						Min.	**Max.**
Step 1	Trust	.565	.388	.145	1.760	.822	3.766

	Openness	.582	.415	.160	1.790	.794	4.035
	Collaboration	.209	.256	.415	1.232	.746	2.037
	Constant	−4.503	1.087	.000	.011		

Belief ratio test: $\chi^2 = 125.911$, $df = 3$, $p < 0.001$.
Hosmer-Lemeshow test: $\chi^2 = 17.595$, $df = 7$, $p < 0.014$.
Nagelkerke $R^2 = 0.33$.

Table 4. Results of binary logistic regression forecasting the timeliness of adopting the budget based on trust.

Variable		Score B	Standard error assessment	p	Odds ratio Exp(B)	95% confidence interval for Exp(B)	
						Min.	Min.
Step 1	Trust	1.195	.261	.000	3.305	1.981	5.512
	Constant	−4.147	1.045	.000	.016		

Belief ratio test: $\chi2 = 130.567$, $df = 1$, $p < 0.001$.
Hosmer-Lemeshow test: $\chi2 = 5.921$, $df = 7$, $p < 0.052$.
Nagelkerke R2 = 0.29.

Table 5. Contingency table (independent variable: trust).

Actual value			Forecast value		
			Timely adoption of the budget		**Percentage of correct**
			No	Yes	
Step 1	Timely adoption of the budget	No	21	23	47.7
		Yes	11	67	85.9
	Total percentage				72.1

Having determined that using a model with all three scores for communication quality was not appropriate, we conducted separate binary logistic regressions with separate variables.

Trust has a statistically significant effect on the timely adoption of the budget.

Based on the Hosmer-Lemeshow test ($\chi^2(2) = 5.921$, $p = 0.052$), we have concluded that this was an appropriate model (**Table 4**). The model explains the 29% (Nagelkerke R^2) variance in the timely adoption of the budget and has correctly classified 72.1% of the examples (**Table 5**), while trust is in direct proportion to the probability of the budget being adopted on time.

Openness has a statistically significant effect on the timely adoption of the budget. Based on the Hosmer-Lemeshow test ($\chi^2(2) = 11.12$, $p = 0.004$), we have concluded that this was not an appropriate model (**Table 6**).

Table 6. Results of binary logistic regression forecasting the timeliness of adopting the budget based on openness.

Variable	Score B	Standard error assessment	p	Odds ratio Exp(B)	95% confidence interval for EXP(B)	
					Min.	Min.
Openness	1.173	.250	.000	3.230	1.978	5.276
Constant	−3.759	.933	.000	.023		

Belief ratio test: $\chi^2 = 128.825$, $df = 1$, $p < 0.001$.
Hosmer-Lemeshow test: $\chi^2 = 11.12$, $df = 2$, $p < 0.004$.
Nagelkerke $R^2 = 0.305$.

Table 7. Results of binary logistic regression forecasting the timeliness of adopting the budget based on collaboration.

Variable	Score B	Standard error assessment	p	Odds ratio Exp(B)	95% confidence interval for Exp(B)	
					Min.	Min.
Collaboration	.774	.187	.000	2.169	1.504	3.129
Constant	−1.981	.635	.002	.138		

Belief ratio test: $\chi^2 = 138.812$, $df = 1$, $p < 0.001$.
Hosmer-Lemeshow test: $\chi^2 = 7.508$, $df = 2$, $p < 0.057$.
Nagelkerke $R^2 = 0.214$.

Table 8. Contingency table (independent variable: Collaboration).

Actual value			Forecast value		
			Timely adoption of the budget		Percentage of correct
			No	Yes	
Step 1	Timely adoption of the budget	No	15	29	34.1
		Yes	12	66	84.6
	Total percentage				66.4

Collaboration has a statistically significant effect on the timely adoption of the budget. Based on the Hosmer-Lemeshow test ($\chi^2(2) = 7.508$, p = 0.057), we have concluded that this was an appropriate model (**Table 7**). The model explains the 21.4% (Nagelkerke R2) variance in the timely adoption of the budget and has correctly classified 66.4% of the examples (**Table 8**). Collaboration is in direct proportion to the probability of the budget being adopted on time.

We confirm hypothesis 1 that trust, openness, and daily collaboration between the mayor and finance director are important factors that affect the timely adoption of the budget.

4. Survey results and discussion: the clarity of information the mayor presents to the municipal council

We tested hypothesis 2 that the clarity of information the mayor presents to the council is an important factor in the timely adoption of the local community's budget by testing the difference between the arithmetic mean for two independent samples. Finance directors are present at council sessions when budgets are being adopted and have insight in the clarity of information exchanged between the mayor and the municipal council. We asked finance directors to rate the clarity of information provided by the mayor to the municipal council. They rated the clarity of information on a 5-point Likert scale, with 1 meaning inadequate and 5 excellent.

Table 9. The effect of the clarity of information the mayor presents to the municipal council on the timely adoption of the budget.

	N						
Timely adoption of the budget	**In force**	**Missing**	**Average value**	**Median**	**Standard deviation**	**Minimum**	**Maximum**
Yes	78	0	3.73	4.00	.976	1	5
No	44	0	3.07	3.00	.789	2	5

Table 10. Mann–Whitney U for testing the effect of the clarity of information the mayor presents to the municipal council on the timely adoption of the budget.

Rate the clarity of information the mayor presented to the municipal council	Timely adoption of the budget	N	Average range	Sum of ranges	Mann–Whitney U	p
	No	44	46.03	2025.50	1035.50	0.000
	Yes	78	70.22	5477.50		
	Total	122				

We once again applied the Mann–Whitney U test to check whether the score for the clarity of information provided to the municipal council was statistically significantly higher in local communities where the budget was adopted on time than in the local communities where it was not.

The average score for clarity of information provided by the mayor to the municipal council in local communities where the budget was not adopted on time was 3.07 ± 0.789 (Me = 3). The average score for clarity of information provided by the mayor to the municipal council in local communities where the budget was adopted on time was 3.73 ± 0.976 (Me = 4) (**Table 9**).

We determined that the clarity of information the mayor presents to the municipal council has a statistically significant effect on the timely adoption of the budget ($U = 1035.50$; $p < 0.001$) (**Table 10**). The clarity of information provided by the mayor to the municipal council in local communities where the budget was not adopted on time was lower.

We subsequently accept hypothesis 2: The clarity of information the mayor presents to the council is an important factor in the timely adoption of the local community's budget.

Fifty percent of the respondents from municipalities where the budget was not adopted before 1 January 2014 rated the clarity of information with 3 or 4.

5. Conclusion

Other findings indicate that "the global financial crisis has had an impact on Local Government forcing it to be more transparent in management of public resources" [13]. For this, collaboration between the mayor and finance officer and between the mayor and council is necessary. Here there are summarized results of our empirical research. As we did not find exactly the same international empirical research, we will show the results of those which are partially similar in content. With our research, we have thus filled the research gap both empirically and theoretically.

We found out that the collaboration between the mayor and finance director is an important institutional factor affecting the adoption of the budget, with mutual trust, openness, and daily collaboration as the basis for good and timely performance of tasks and preparation and adoption of the local community's budget. Other research showed that local governments, representatives, and citizens often perceive public budgeting as a difficult administrative task that should be handled by experts who have specialized knowledge, technical skills, and experience [17]; we assumed that these experts are finance directors. In our research we made a step forward and claimed that the collaboration between the mayor and finance expert in preparing the budget is necessary. Other research shows that a wise council and mayor, as the first member of the council, would not want to act without receiving a kind of advice that these experts can provide [18]. Another research shows that as the first citizen of a municipal area, the mayor is entrusted with varying roles, also with financial roles where the trust and collaboration between the mayor and finance officer is important [19]. The results of empirical research also indicate that only about 50% of managers actively participate in the budget preparation process and 67% of finance officials have very good working relations with their managers [19].

Good collaboration between the mayor and municipal council and the clarity of information the mayor presents to the council are also important for local community's effective work. It is very important because findings indicate "city councils have the right to reject a mayor's proposed budget". At the municipality level, other similar findings indicate that political variables are those factors that play the most important role [20]. Another similar empirical research also shows that through citizen collaboration in the budget process, however, the distance between principals and agents became somewhat smaller, and therefore it is more likely that the budget will be adopted [21]. Also [22] argues that various cities show decreasing participation over the past years; their research with analysis of interviews with employees of public administration identified 12 individual barriers for low participation in the local budgeting. Among these barriers clarity of information was not presented by the mayor to municipal council (as representative of citizens).

This was upheld by the analysis which showed that the clarity of information presented to the councilors by the mayor affected their decision-making and the timely adoption of the budget.

Based on the analysis findings and the empirically proven impact that separate factors have on the adoption of the local community's budget, we recommend:

- ❖ The municipal administration and finance directors should ensure all activities are conducted on time.
- ❖ The mayor should daily communicate with the municipal administration and finance directors about the activities in compiling the materials for the local community's budget, since daily communication helps grow mutual trust and vice versa.
- ❖ The mayor should resolve any confusion in drawing up the budget with knowledge and experience and together with municipal administration and finance directors.
- ❖ Timely adoption of the budget document, in which the municipal council adopts the revenue and expenditure plan for the year. The budget must be adopted before the period to which it applies:
- ❖ Being realistic in budget planning
- ❖ Drawing up a transparent, accurate, clear, and understandable budget
- ❖ Following the principles of economical and efficient operations and looking for cost-effective solutions
- ❖ Complying with the law, all the revenue and expenditure must be based on the law, since municipalities can only do what is stipulated by the law
- ❖ The municipal administration's body in charge of the finance must, after receiv- ing basic economic premises and assumptions for drawing up the draft state budget in accordance with Article 18 of ZJF, send instructions for budgeting to all municipality budget's direct users in time, providing the following information:
- ❖ Basic economic premises and assumptions for drawing up the state budget draft
- ❖ Description of the local community's planned policies
- ❖ An estimation for the revenue and expenditure account, the financial receivables and investments account, and the budget financing account for the following 2 years
- ❖ A draft financial plan proposal for every direct budget user for the following 2 years
- ❖ The method of drawing up the municipal budget and proposed financial plans for direct users and a schedule
- ❖ The method of drawing up development and hiring plans, as well as procurement and construction plans of direct users, and the deadlines for submitting
- ❖ The method of drawing up the plan of selling municipal tangible assets and the deadlines for its submission
- ❖ The method of drawing up financial plans (separately for revenue and expenditure), development plans and explanations for direct users, and the deadline by which direct users must submit proposed financial plans, as well as development plans and explanations to the municipal administration body in charge of finance

We see an opportunity for further research primarily in looking for other factors that affect the timely adoption of local community budgets, which could include the number of municipal administration employees, the way the administration is organized, the municipal council's political structure, and the mayors' political affiliation. The survey should also include mayors, municipal administration directors, municipal councilors, and the Ministry of Finance. Further research should theoretically and empirically prove that other factors also importantly affect the timely adoption of the local community budget and subsequently the quality and scope of tasks that are performed in the local community [23].

The main limit of this research is that it was focused on the finance director's perspective while not including the opinions of other stakeholders, especially the mayor and the municipal council.

References

1. Frolova EV, Vinichenko VV, Kirillov AV, Rogach OV, Kabanova EE. Development of social infrastructure in the management practices of local authorities: Trends and factors. International Journal of Environmental & Science Education. 2016;**11**(15):7421-7430
2. Setnikar Cankar S. NEXT4PA— Kreativnost in inovativnost v javni upravi. In: Setnikar Cankar S et al., editors. Kreativnost in inovativnost za javno upravo: NEXT4PA—Kreativna javna uprava. Ljubljana: Faculty of Public Administration; 2014. pp. 14-23
3. Cvikl M, Korpič Horvat E. Proračunsko Pravo: Priprava, izvrševanje in Nadzor proračunov. Vol. 373. Ljubljana: GV Založba; 2007
4. George J, John S. Local government: The past, the present and the future. Public Policy and Administration. 2012;**27**(4):346-367
5. Copus C. British local government: A case for a new constitutional settlement. Public Policy and Administration. 2006;**21**(2):4-21
6. Pernek F, Kostanjevec B, Čampa M. Davčno Pravo in Javne Finance. Novo Mesto: Faculty of Business and Management Sciences; 2009. p. 133
7. McIlquham-Schmidt A. Strategic Planning and Corporate Performance. What is the Relationship? Aarhus: School of Business at Aarhus University; 2010
8. Hopkins WE, Hopkins S. Strategic planning—financial performance relationships in banks: A causal examination. Strategic Management Journal. 1997;**18**(8):653-652
9. McKiernan P, Morris C. Strategic planning and financial performance in UK SMEs: Does formality matter? British Journal of Management. 1994;**5**(1):31-41
10. Schack RW. Political, Institutional and Organizational Effects on the Accuracy of Local Government Revenue Estimation. Connecticut: University of Connecticut; 2000
11. Žohar F. Participativno Upravljanje občine—Sodelovanje Pri Pripravi proračuna. Vol. 5. Zlati Kamen: Revija Za Lokalno Samoupravo; 2011. pp. 49-51
12. Grossi G, Reichard C, Ruggiero P. Appropriateness and use of performance information in the budgeting process of municipalities: Some experiences from Germany and Italy. Public Performance & Management Review. 2016;**39**(3):581-606
13. Tejedo-Romero F, Ferraz JF, de Araujo E. Determinants of local governments' transparency in times of crisis: Evidence from municipality-level panel data. Public Performance & Management Review. 2016;**39**(3):581-606

14. Jørgensen O. Place and city branding in Danish municipalities with focus on political involvement and leadership. Place Branding and Public Diplomacy. 2016;**12**(1):68

15. Mouritzen PE, Svara JH. Leadership at the Apex: Politicians and Administrators in Western Local Governments. Pittsburgh: University of Pittsburgh; 2002

16. Haček M, Kukovič S, Grabner A. Odnosi Med Organi Oblasti v Slovenski Lokalni Samoupravi. Available from: http://druzboslovne-razprave.org/stevilka/2013/72/ [Accessed: 2.3.2016]

17. Švaljek S, Rašić Bakarić I, Sumpor M. Citizens and the city: The case for participatory budgeting in the city of Zagreb. Public Sector Economics. 2019;**43**(1):21-48. DOI: 10.3326/pse.43.1.4

18. Fenn M, Siegel D. The evolving role of city managers and chief administrative officers. IMFG Papers on Municipal Finance and Governance. 2017;**31**:1-30

19. Mahlaku BT. Chapter 1. Introduction: Orientation, Problem Statement, Research Method. Available from: http://dspace.nwu. ac.za/bitstream/handle/10394/13176/ Mahlaku_BT_Chapter_1-5_ Bibligraphy%26Appendices. pdf?sequence=2 [Accessed: 11.11.2018]

20. Ott K, Mačkić V, Bronić M. Budget transparency of local governments: The political economy of city and municipality budgets in Croatia. Economic Research. 2018;**27**(4):629-647. DOI: 10.5559/di.27.4.03

21. Kim S, Schachter H. Citizen participation in the budget process and local government accountability: Case studies of organizational learning from the United States and South Korea. Public Performance & Management Review. 2013;(36):456-471. DOI: 10.2307/23484750

22. Zepic R, Dapp M, Krcmar H. Reasons for low participation in German participatory budgeting: A public administration perspective. In: Proceedings of the 17th European Conference on Digital Government. 2017. Available from: https://www.researchgate.net/ publication/317571030_Reasons_ for_low_Participation_in_German_ Participatory_Budgeting_A_Public_ Administration_Perspective/download [Accessed: 1.8.2018]

23. Gaber Sivka N. Dejavniki Sprejetja Proračuna Lokalne Skupnosti. Koper: Faculty of Management; 2016. p. 85

Section 2
Urban Planning and Regeneration

Chapter-3

Symmetrical Aspects of Urban Regeneration in Seoul

1. Introduction

Urbanization has proceeded rapidly since the industrial revolution, and explosive expansion of the city brought about suburbanization to avoid high population density, traffic congestion, and contaminated environment. New towns were constructed outside the cities of industrially advanced countries. In Korea, there were three phases of constructing new towns since 1980s, and shortcomings of fast growth have been exposed.

Because new towns functioned as bed towns, they had comprehensible problems. Therefore, people began to return to old towns, and urban remodeling was needed to improve out-of-dated infrastructure and weary environmental conditions. In doing so, local characteristics were lost, and uniform scenery became common. Urban regeneration aroused as a countermeasure to this mishap.

In this study, urban regeneration reinforced with smart technologies is suggested to revive lost placeness, communal connectivity, and urban orientation. Gentrification is another important issue to be resolved for the sustainable urbanization in old towns. Symmetrical aspects were main topics to be dealt with for the successful urban regeneration.

2. Need for urban regeneration

2.1. New towns

In order to meet the needs of modern urban functions which had begun with the industrial revolution, the concept of new town should have resolved the exposed problems therewith. It started as building of "new regional community" in England and executed the recovery projects worldwide after World War 2. In Korea, the new town projects were initiated in 1960s to attain both the goals of national and regional development and resolutions of urban problems in mega cities. The construction of new towns aims at stabilizing the housing market in metropolitan area and also solving dwelling problems.

By 1980s, inside Seoul, there is no more land for housing projects, and so outside the green belts, new towns were built. Five new towns under comprehensive plans accommodate offices, housings, commercial buildings, municipal boroughs, sports facilities, and parks. By 1990s, the dissemination of new housing helps stabilizing housing costs, and the extension of roads and subways improves transportation system. However, vicinity towns of smaller sizes were developed with poor plan and consequently caused serious problems due to insufficient infrastructure and coarse conditions.

Besides, local employment did not match the migrating population to new towns. Thus, new towns became bed towns for metropolitan.

Twelve new towns were planned after second millennium to stabilize land supply for comfortable housing and to establish foothold for economic self-reliance. They were expected to loosen the stress of overcrowded metropolitan life and to facilitate secured dwelling. Unlike the new town projects of 1980s, these new towns faced the unpopularity for sale. It is due to the supply overflows in vicinities, insufficient financial investment for housing, and poor infrastructure conditions.

Most recent attempt for new towns targeted the area between Seoul metropolitan area and first-phase new towns (Figure 1). Korean government boosted housing projects for the economic growth, but the market responded in a different way only to raise existing housing prices. The number of empty housing units increased, and in return, social problems have been occurring in recent new towns. New town projects also have led to the speculation in real estate and raised land prices. The emphasis on housing policy without consideration of urban structure brought about social inefficiency and lost urban characteristics (placeness, connectivity, and orientation).

New towns around Seoul expand in every direction with more than 40 km radius. They are not differentiated in their functional divergence, but just bed towns, so to speak, and more new town plans are in discussion. Spatially, dispersed new towns have serious problems in public transition and traffic congestion. Present megalopolis traffic networks cannot handle explosive increase of commuters between new towns and the city.

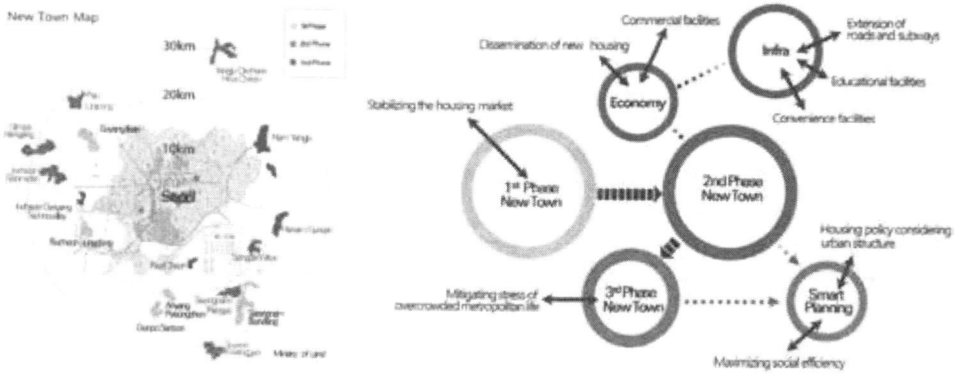

Figure 1. New town map and planning policy.

The present process of traffic network building after site development will weaken the traffic control system and ease the transportation for commuters. With the repetition of current processes without understanding the relationship between the spatial structure and traffic networks, it is almost impossible to realize the smooth and effective traffic networking. Comprehensive planning and execution of traffic networks in megalopolis regions are required to improve traffic conditions fundamentally.

2.2. Urban remodeling vs. urban regeneration

In the last 50 years, Korean government's urban development plan has centered on the expansion of new towns and new neighborhoods because of rapid industrialization and urbanization. The result is the gravitation of population in major cities and disorganized ordination in vicinities. Therefore, the urban remodeling projects were introduced for the remedy. Unexpectedly, it dropped life conditions and lost historical and cultural identity.

On the other hand, suburbanization of modern cities raised the land price of suburbs with popularity, while in the old town center, the population reduced and the physical conditions antiquated. Worn out buildings were torn down, and the old town was remodeled. This method with policy caused various urban problems from environmental dilapidation and ecological incongruity.

About 10 years ago, the development of new towns advocating smart city surfaced as the nucleus of urban growth. But this trend of urban growth deepened the extremity of old and new town's circumstances. In the process of separate development of new towns, city identity is very likely to be diminished and eventually lost. Against this deformity of urban growth introduced is the urban regeneration scheme and plan.

In result, the reduced population growth, change of industrial structures, unplanned urban expansion, and aging residential environments caused urban fall-off incrementally. Urban regeneration attempts to rejuvenate degenerating cities by enforcing regional competency, inducing new functions, and utilizing local resources economically, socially, and environmentally. Contralto urban maintenance, urban regeneration takes higher priority in reserving sustainable community integrating housing policy with socioeconomic perspectives.

3. Shortcomings of urban regeneration

3.1. Placelessness

Since 1970s of new urbanism era, the connection between space and people and its concept has received growing interest. According to Relph [1], place consists of physical condition, functional behavior, and meaning, and placeness arises from the interactions between them. Schulz [2] stated place was the collection of existential objectives more than space in location, and it is due to the interactions between people and their settings. In 1981, Steele [3] announced that placeness meant individuals' specific experience upon specific environment, which was human reaction to stimulus from environmental settings. It is the interactive concept, which includes feelings, consciousness, behaviors, and their interactions inside its existence.

Place gives people the sense of belonging or connecting to that place. However, regional civic authorities in Korea for recent urban readjustment plans have ignored the unique character of neighborhood community and the continuity of time. The emphasis is mainly laid on solving functional problems due to the high population density and worsening environmental conditions since the rapid industrialization from 1970s. Therefore, urban scenery appears uniform in most urban sectors miss- ing diversity and identities. It was named placelessness by Relph.

3.2. Disconnectedness

Urban spaces consist of both artificial elements such as buildings and bridges and natural elements such as woods and rivers not to mention people living in there. These elements have certain disposition and pattern by bilateral interactions so that they impose urban characteristics and consistent schemes. Together, they integrate urban spaces resulting in enhanced functional efficiency, environmental balance, bestowed placeness, and elevated spatial potentiality.

Undoubtedly, urban connectivity implies organic connection of these elements which covers not only the physical link between them but also the concoction of consisting elements of urban spaces. The flow path of pedestrians and cars, mixed use development, density and shape of facilities, energy efficiency, landscape, landmark and significant viewpoints, and block/section division are all critical components of the urban connectivity.

Spatial interactions instigating connectivity occur at the boundaries by which spaces are divided by the different land uses, and they include contact areas and their surroundings. Natural parks will immediately appear as direct interactions between urban spaces and represent visual connectivity. The mental connectivity of accumulated time with respective spaces is required in existing cities, especially in urban regeneration.

3.3. Disruption of social community

Human activities and their domains are fundamental aspects of urban life to maintain the sustainability of cities. In precedence of environmental issue, the connectivity of neighbors is the core of urban function. Human relationship and communication compose the connectivity of diverse classes of citizens. It is an urban community where these actions are barreled.

In Korea, autogenous local community had not been formed until the rebirth of local autonomy governance, since the urbanization occurred in short period of rapid industrialization. Japanese occupation and Korean War led to the disconnection of historical tradition to urban culture, and urban community did not take root in sudden social transition. Community was formed by the authoritative government and was functioned as the mass mobilization for public intentions.

Urban community is meant to be a city itself or a number of communities that are scattered over the city. Until recent period, urban policy seems to have been governed by the market throughout the world. The counteractive issue of community building is raised quite recently. The community building starts with the protection of placeness.

In Korea, rapid urbanization dismantled rural communities, and urban communities have not been completed up to appropriate level. Lagging and underdeveloped urban areas do not get satisfactory effects by the government support or market-led approaches. Most regional cities just tried to resemble Seoul, the capital city of administration, economy, and culture, in which they have lost locational characteristics and uniqueness of place. Identical urban scenery and culture are brought out.

Now civil movements are trying to rebuild communities to resolve housing problems and to improve environmental conditions such as green housing collaboration, urban village alliance, and livelihood joint union. These communities acknowledge that it is not possible for government or market alone to resolve the conflicts in transit, housing, environment, energy, education, security, and well-being. Government does not cope with the versatile lifestyles and living demands, and the market only deals citizens as consumers and not as dwellers. Urban regeneration is getting spotlight from the perspective of organizing local communities, vitalizing local economy, and preserving historical valuables.

3.4. Gentrification

Gentrification refers to the phenomenon that residents or leaseholders being kicked out when the property value goes up according to the change of land use. In the course of re-urbanization, cities may confront gentrification situation. Urban development or maintenance projects reform physical space of the corresponding areas, and at the same, they change the context and placeness therewith. Reformation and reorganization revitalize the area with new functions and attractions and eventually lift real estate value of that area. It then invites developers and rich people to invest in the area, and famous stores and franchise business move in. Once underdeveloped, neighborhood with destitute artists, unrestrained bohemians, and small manual businesses turns into vital and glittering spot with visitors and shoppers. But original residents and leaseholders have to move out because of soaring rental fees, and the changed neighborhood loses its original charms and identity.

There are three types of gentrification in Seoul. (1) Commercial leading type: redevelopment of commercial area replaces small businesses with franchise business (Figure 2). The fact that franchise business routes out small business raises negative social issue. (2) Cultural initiative type: lagging neighborhood in old town of Seoul had potential for cultural delivery since artists moved for studios with inexpensive rent (Figure 3). Ubiquitous demand for modern cultural consumption commercialized the area. (3) Cultural tourism type: commercial area with attractive touring places draws large amounts of visitors (Figure 4). The same issue arouses as type 1.

Figure 2. Commercial leading type 1 (photos, Korea Tourism Organization).

Figure 3. Cultural initiative type (photos, Korea Tourism Organization).

Figure 4. Cultural initiative type (photos, Korea Tourism Organization).

3.5. Smart city schemes

How does a city become smart? Smart city is not just a platform for computer users, but an integrated system merged into everyday life. Cities have thrived with the progress of industrialization, so functions and convenience are aptly emphasized. Ongoing development of smart systems looks for eminent security, comfortable service, and technological advance. It shapes cities and changes the society.

A number of international groups and adjoining cities have experimented smart city projects. They focused on the solutions for realistic and functional problems of contemporary cities. Traffic scheduling/control, intelligent CCTV's, energy efficiency, and environmental improvement are among them, and recently, smarter solutions are being seeked by use of big data, IoT, AI, and block chain technologies. A smart city is the platform of the fourth industrial revolution, which claims to change our life paradigm (Tables 1 and 2).

Table 1. Examples of application of smart technology (Europe).

Amsterdam, Netherlands	Kalasatama, Helsinki, Finland	Columbus, USA	London, UK
• 5G, Sensor, Camera, Open access data, Platform • Organization 3.0, (Neighbor collaboration) • "Cross-silo" collaboration • Digital connectivity • Network city	• "Sohjoa" autonomous electric car • "MaaS", Mobility as a Service • Whim service • Smart Grid • Agile Piloting	• "Virtual Singapore" platform: safety system • Smart City Challenge • Traffic, logistics • Medical, Travel smart app • Mobility and connected • Autonomous driving	• API traffic information • Rent public bicycle • Location Information • City Data Analytics Program • Borough Data Partnership • Economic Fairness—utilize metrics

On-offline platform, public-private cooperation	Government, developer, neighbor, civic group, scholar co-create system (Innovator's club)	Data integration with other cities	Data integration

Table 1.
Examples of application of smart technology (Europe).

Now, the focus of smart city projects moves on into the human network and the connection between various classes. Three components are the key value-driven targets for future smart cities: people, environment, and technology. People's happiness is the ultimate goal of the smart city research and development in which urban space should be environmentally safe and socially connected with help of advanced technologies. To make the old town smart is a tough challenge because the infrastructure is old and urban structure is irregular. To overcome this challenge, symmetrical aspects of urban regeneration should be examined and key solutions should be developed.

Table 2. Examples of application of smart technology (Asia).

Hángzhōu, China	Singapore	Sejong, South Korea
• Paperless • IOT and Digital wallet, Block-chain, Intelligence information technology, smart-payment (Ali-pay), Mobile payment system • Tao café face recognition, QRcode, scanning unattended payment system.	• Living Lab • Virtual-reality city platform, Digital twin, Block-chain • Intelligent platform	• Mobility: Smart Street, • Smart Light • Health care • Education • Energy and Environment: smart road lighting system, Smart Park, Kiosk, Smart-poll service (emergency safe bell)
	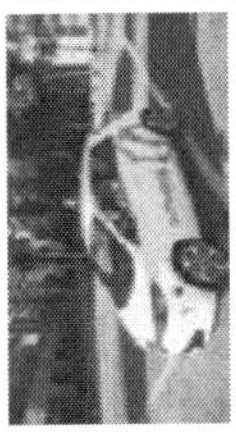	
Intelligence information technology	Intelligent platform	Data integration

Table 2.
Examples of application of smart technology (Asia).

4. Symmetrical aspects of urban regeneration

In this paper, symmetry indicates the complete status of intended urban regeneration, and symmetrical aspects point the composing factors for the outcome. There are three categories in those aspects, (1) the harmony of people, environment and technology, (2) the revival of placeness, connectivity and urban orientation, and (3) the integration of social community and economic community.

City is the test bed for embodying human life and smart technologies. Intelligent information management by big data and AI technologies would integrate traffic control system, crime surveillance system, energy management system, and water service system with the purpose of convenient and safe urban living. In accordance with several European cities, Korea initiated smart living lab projects that are open and cooperative to an innovative model.

It is difficult to apply smart city scheme to old town regeneration because the old infrastructure is not often compatible with smart technology. Nevertheless, smart parking system, smart garbage collection system, self-supporting energy system, and secure footpath system have been undergone as fourth industrial revolution technology applications. Undoubtedly, these attempts will increase the value index of urban life. However, the harmonious progress of people, environment, and technology can never be overemphasized in smart city projects.

Three keywords are derived for urban regeneration: living (placeness), social ecology (connectivity), and sharing economy (urban orientation).

4.1. The revival of placeness

Sircus [4] said successful place made people focus on the purpose of place and provide them with well-composed story for emotional experience. Good place is filled with vitality and vigor, and thus, Jacobs [5] suggested streets to be given back to citizens. The project for public space [6] reported essential factors for the successful place should be equipped with comfort, visual friendliness, accessibility, connectivity, utility, and intimacy. Montgomery [7] elucidated the success of public spaces entirely depends on establishing the basis for the exchange of diverse classes of people and culture. That will enable space to become meaningful and attractive place.

Seoul is pretty old city and owns significant amount of time trace marks all over the city. However, urban remodeling and reconstruction erased them not to mention the damage from Korean War. Therefore, smart technology could be the answer to restore the memory. History telling by smart devices with virtual reality can be a meaningful experience on specific places. Every city has topological peculiarity, and it used be a natural and rational route to navigate the city.

Disappeared nodes, edges, and boundaries [8] are revived as familiar and attractive places virtually (Figure 5).

4.2. The revival of connectivity

Natural resources like lakes and rivers are well accepted as connecting places for residents and visitors. Smart technologies will not only improve their environmental conditions but also enable people with diverse interests to participate in more activities simultaneously. People could gather easily and safely and meet others with intentions. The extent of connectivity will increase significantly.

In addition, recent trends of sharing housing and transportation are also encouraging the meeting of people. This is made possible in the first place by the development of social network for individuals using IT and communication services. Sharing public data is another factor to encourage people to organize groups of similar social interests. Collecting big data and analyzing them could intrigue people and organize special interest groups and locate them online urban hotspots (Figure 6).

Figure 5. The revival of placeness.

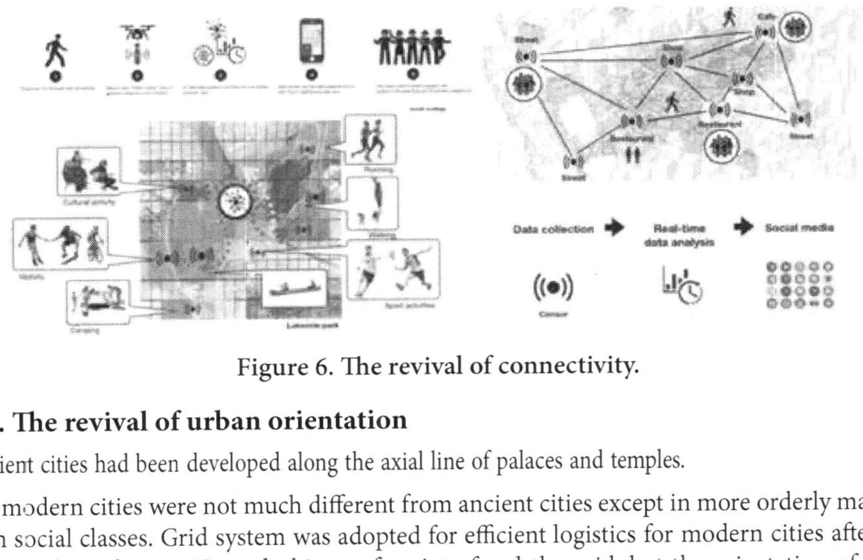

Figure 6. The revival of connectivity.

4.3. The revival of urban orientation

Ancient cities had been developed along the axial line of palaces and temples.

Pre modern cities were not much different from ancient cities except in more orderly manner with social classes. Grid system was adopted for efficient logistics for modern cities after the industrial revolution. Natural objects often interfered the grid, but the orientation of cities was kept with political and social hierarchies. As modernization advances, financial district became major point in the city. Cities have been shaped with landmarks, boundaries, nodes, and edges.

In old towns, urban remodeling broke the previous order and changed the urban orientation. Accumulated traces of the past are erased, and strange new blocks appeared in a brief time. Individual stores were replaced by large shopping malls. Residence areas became condominium sites. Besides, returning population to the old town led to gentrification. Familiar sceneries of small shops and alleys with neighbors were hardly seen. Urban regeneration should achieve the restoration of neighborhood and urban orientation (Figure 7).

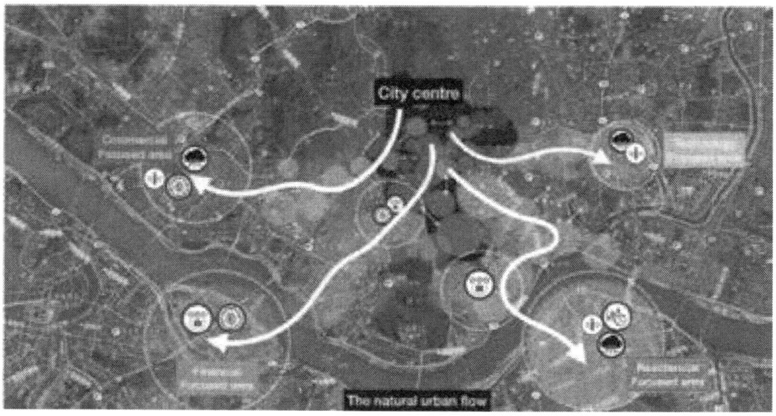

Figure 7. The revival of urban orientation.

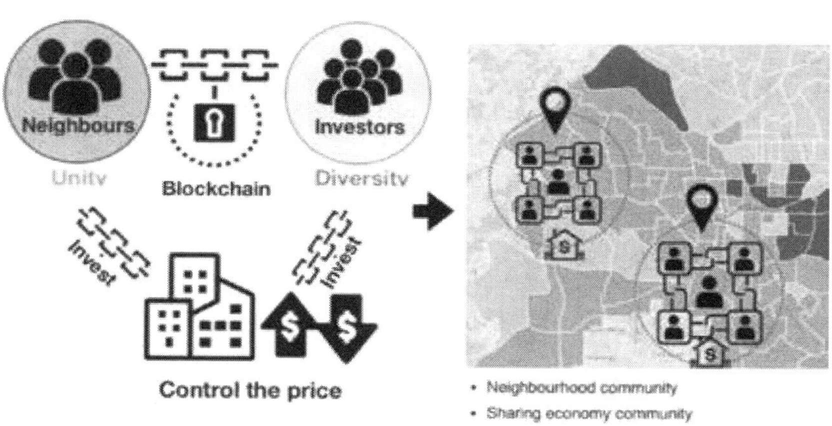

Figure 8. Ecologically integrated community.

Final mission of urban regeneration is to avoid gentrification problems because it reorients artificially the natural urban flow. Residents and small business owners live in the same neighborhood may be able to keep unity as one half of symmetrical aspects. Using block-chain technology, anonymous investors can join the neighborhood, and they will provide diverse ingredients as the other half of symmetrical aspects (Figure 8).

5. Conclusions

Defining symmetrical aspects of urban regeneration in this study suggested the experimental smart service for the revival of placeness, connectivity, and urban orientation. The service measures neighborhood alleys and small piazzas in scale, density, absorption, and deterioration with digital devices and analyzes collected data for familiarity and network tension with big data and AI technologies.

Autonomous protection and control of back alleys and unplanned street corners would provide residents safety and comfort. Finding viewpoint among the area would give an idea where to place a new piazza and to induce sustainable urban orientation.

The ecologically integrated community will help eliminate gentrification phe- nomenon and sustain urban layout and social order. Ecological integration requires the fusion of dissimilar communities, which are defined by geographical proximity and by sharing economy. Together, they will contribute to keep the community sustainable.

References

1. Relph E. Place and Placelessness. London: Pion; 1976
2. Schulz. Architecture: Meaning and Place, selected essays. Newyork: Rizzoli; 1986
3. Steele F. The sense of place. Boston: CBI Pub Co; 1981
4. Sircus J. Invented places. Prospect. 2001;81:30-35
5. Jacobs J. The Death and Life of Great American Cities: The Failure of Modern Town Planning. New York: Random house; 1961
6. Meeyoung K. A study on the new possibility of public space in contemporary cities. Journal of the Korean Geographical Society. 2017;14: 455-463. DOI: 10.25202/JAKG.6.3.11
7. Montgomery J. Making a City: Urbanity, vitality and urban design. Journal of Urban Design. 1998;3:93-116. DOI: 10.1080/13574809808724418
8. Lynch K. The image of the city. The MIT Press; 1960

Chapter-4

Application of a Metabolic Thinking Driven Sustainability Framework in Early-Stage Planning of Eco-City

1. Introduction

The fast urbanization rate together with increasing population and consumption is challenging the long-term sustainability of our social systems and supporting ecosystems on earth. Without healthy ecosystems, there can be no sustainable urban systems. The signs of instability can be seen in environmental degradation, e.g., climate change and loss of biodiversity. Also, the increasing use of materials and energy creates competition and international conflicts. It is important to understand that there is a causal link between the increasing use of resources and negative effects on environment. This is valid both on local and global levels. With a growing population on earth, a central question is if urbanization is the solution or a threat to sustainable development? Although cities cover a small part of the world, their physical and ecological footprints are much larger. Cities produce more than 70% of the world's greenhouse gas (GHG) emissions and use 80% of the world's energy. On the other hand, cities are regarded as centers for innovation and they account for 82% of global GDP in 2014, and by 2025 this will rise to an estimated 88%. There will be 230 new cities by then, all in middle-income countries. One hundred cities in China alone may account for 30% of global GDP at that time [1]. From these figures, it is clear that new ways have to be found to plan cities in order to achieve a better balance between growth and resource consumption. This is especially important for developing countries where also population growth is high.

Different concepts and solutions for sustainable urban development have been presented, but the solutions seem inadequate. Emissions and loss of biodiversity continue to increase. Proposed solutions are mostly related to the idea of ecological modernization which essentially proposes that we can actually adapt our urban metabolism in line with sustainable development just by applying new technologies and new and smarter systems solutions. In this way it is assumed that we can keep the existing consumption patterns, and developing countries can go the same way the developed countries have done. The ideas of ecological modernization have been questioned in several papers [2], but it is an agenda which is favorable to use for politicians. It will raise no fundamental conflicts with their voters who are looking for a decent life with high living standard especially in the developing countries. The solution proposed is often to develop smarter cities. However, all ideas around this concept are vague and scattered [3].

The central argument, which will be developed in this chapter, is that sustain- able urban development is a normative process and must involve local actors including people living in the area in order to go from awareness to action. We will also discuss role models and principles for doing this in order to reduce resource consumption in urban development.

2. Aims and objectives of this chapter

The general aim of this chapter is to outline why we have ended up in a situation with fast urbanization and increasing resource consumption and to find a starting point for going from awareness to action toward more resource-efficient urban systems.

The objectives in the chapter are:

- Give a historical background for the growing resource consumption and related environmental impacts in urban development. How did we end up in the current situation?
- Discuss the main driving forces for increasing resource consumption and environmental effects linked to urban development.
- Discuss if we find a foothold for objective decisions on sustainable development based on environmental considerations.
- Outline support for planning of more resource-efficient urban systems based on a case study in China using a recent Swedish support system for urban planning.

3. The history of urban development and human resource use

For thousands of years, human settlements existed within the biosphere without affecting it more than locally. The biosphere includes the regions of the surface, atmosphere, and hydrosphere of the earth occupied by living organisms. Most material and energy needed by humans came from local sources, and waste, mainly organic, was returned locally.

During the last couple of hundred years, a fundamental and dramatic change occurred. The large-scale exploitation of fossil energy and the technical inventions like heat engines allowed a rapid liberation from local energy sources with low intensities like renewable biomass, wind, and streaming water. Also new ways for transportation facilitated the exchange of goods and people on large distances. The starting point for this was the industrial revolution in England about 150 years ago. This resulted in a fast growth of urban settlements. Urbanization then started to form a global network of urban systems and also enabled a fast growth of the global population.

To illustrate this dramatic change, **Figure 1** shows the anthroposphere as a separate entity together with the biosphere and the geosphere. The reason for this is that the large-scale exploitation of energy-rich fossil fuels together with the fast development of technology has created a new situation in the biosphere. Humans are now affecting the environment on a global scale, and the new time periods have been named Anthropocene [2]. From being only local physical flows and stocks, energy and matter are now transported in complex patterns on a global level. The notion of metabolism is used to comprehend all physical flows and stocks of matter and energy within the anthroposphere. Human impact on the environment includes impacts on biophysical environments, biodiversity, and other resources. We here will refer to this as anthropogenic impacts on the environment.

The global metabolism on earth is driven by the energy from the sun. This global system is an open system for energy but a closed system for matter. Matter and energy are flowing between the biosphere, the geosphere, and the anthroposphere. For a long period of time, life has evolved through interactions between the bio- sphere and the geosphere driven by the energy from the sun.

The historically new situation is that humans have increased man-made flow of energy and materials on global level, so the anthroposphere now significantly interacts with the geosphere and the biosphere. This has created to a situation of high complexity, where it is easy to conclude that it is difficult for single nations to change the pattern. Individuals can also feel that their behavior cannot affect the situation. This has led to something like a look-in situation where changing the metabolism is difficult to achieve. The essential question is how and on which level the change process can start. Is it through high-level political decisions or through the change of people's values and behavior on local level?

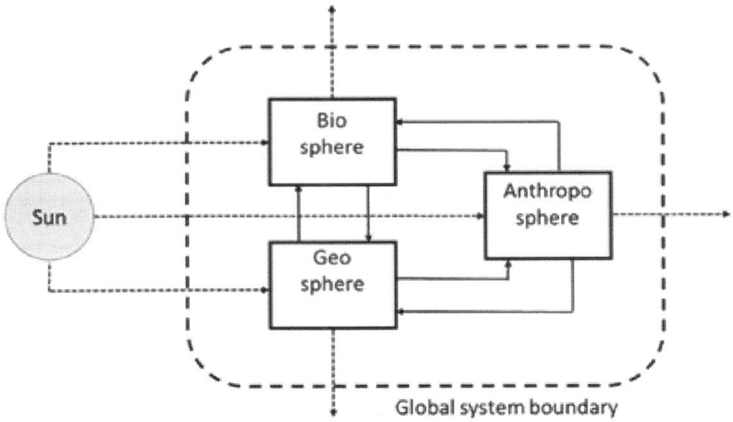

Figure 1. Schematic picture of metabolic processes on earth.

Starting from this perspective, we will look closer into the present and future situation of urban development.

4. Urban development today and in the future

There are several driving forces for urban development and increasing resource consumption in the world today. The most important ones are:

1. A growing population especially in the developing countries.
2. Fast urbanization rate especially in the developing countries and also in the developed countries.
3. Increasing consumption/capita both in developed and developing countries.

A central force for the growing use of resources and environmental impacts is a growing population. The current world population of 7.6 billion is expected to reach 8.6 billion in 2030, 9.8 billion in 2050, and 11.2 billion in 2100, according to a United Nations report [4]. With roughly 83 million people being added to the world's population every year, the upward trend in population size is expected to continue, even assuming that fertility levels will continue to decline. One attempt to describe the role of multiple factors in determining resource consumption and environmental degradation was Equation [5]:

$$I = P * A * T \tag{1}$$

This equation describes the multiplicative contribution of different factors to environmental impact (I) which may be expressed in terms of resource depletion or waste accumulation. The factors are population (P) referring to the size of the human population, affluence (A) referring to the level of consumption by that population, and technology (T) referring to the processes used to obtain resources

and transform them into useful goods and wastes. As stated above, P and A increase on the global level. The only way then to balance an increased impact is then to decrease T. This can be done by producing goods we consume using less energy

and materials. The formula was originally used to emphasize the contribution of a growing global population on the environment, at a time when world population was roughly half of what it is now. It continues to be used with reference to population policy. One should not regard the equation as a simple multiplicative relation- ship among the main factors, but more as a qualitative visualization of the factors. Another way to express this relation is to use gross domestic product (GDP), a measure of industrial and economic activity.

$$\text{Environmental Impact} = \text{Population} \times \frac{\text{GDP}}{\text{Person}} \times \frac{\text{Environmental Impact}}{\text{Unit of GDP}} \tag{2}$$

In the United Nations Sustainable Development Goals, there are 17 goals [6] to handle this complex development, and several of the goals are conflicting. A basic assumption in the UN agenda is that developing countries should have an economic growth of around 7% and that continuing urbanization is important for reducing poverty in the world. China can be a good role model for how economic development has reduced poverty. However, this has been coupled to an enormous increase in resource use and environmental degradation both on local and global levels. Even if we now can produce material goods using less energy and materials, the amount of goods consumed is increasing at a higher rate. Another conflict is that in the UN agenda, it is stated also that democracy is essential for sustainable development.

Sustainable development, it is argued, can only be achieved with the involvement of people in a democratic way since the process involves value statements. At the same time, China is trying to export its growth model and also its political model with central control of one party. One argument put forward is that the developing countries should go through a state of increasing use of resources and environmental degradation before it can turn into a mature developed country. The problem is that the consumption of material goods and energy in developing countries is also increasing and that part of their environmental impacts has been moved to developing countries, e.g., China.

As we can see, the increasing use of resources in the world also increases environmental impacts, e.g., climate change and loss of biodiversity. A central question is if there is a limit for this development or we will reach a tipping point where no return is possible, and do we have time to create change before this tipping point is reached.

5. Is there a red line for our environmental impacts?

The hope behind the UN 2030 SDGs is in line with ecological modernization, that is, economy and ecology can be favorably combined. Technological development and the shift to renewable energy sources can basically overcome the rebound effects. This brings us close to the concepts of weak and strong sustainability where strong sustainability indicates that natural materials and services cannot be duplicated [7]. Already since Thomas Robert Malthus in 1798 published his An Essay on the Principle of Population, it has been discussed if the increase in productivity can keep up with the increase of population on earth, thus avoiding starvation. It is obvious that up till today the increasing productivity in the food sector has kept up with population growth but to the price of larger inputs of nonrenewable sources and environmental degradation. The recent signs of global warming and loss of biodiversity challenge this development and the ideas of ecological modernization. Maybe it is time to realize that the sky is the limit.

The discussions around the future of urban development and solutions for sustainable development and ecological modernization are characterized through many buzzwords which seem to partly replace each other through time when new ones become popular. Examples of these buzzwords are clean technology, circular economy, artificial intelligence, information and communications technology (ICT), big data, etc. These concepts are included in general frameworks for sustainable urban development as low-carbon eco-cities, smart cities, etc. One of the difficulties in handling these buzzwords and concepts is that there is no general definition of sustainability which we can agree upon. Sustainability involves normative aspects, and the only solid boundary we could hope to find is the stability of natural ecosystems that is the fundamental support for all forms of life. In recent years, climate change has become an example of such a red line for non-acceptable environmental impact. The Intergovernmental Panel on Climate Change (IPCC) has formulated a general goal for maximum tolerated global warming to 1.5 degrees. However, there is a high uncertainty in the consequences of what happens if we reach or exceed this goal. Another upcoming focus is the loss of biodiversity. The UN Scientific Expert Panel IPBES recently released the most comprehensive report on biodiversity glob- ally since 2005 [8]. The report states that the protection of biodiversity is crucial for achieving global sustainability goals—and that one million species are threatened with extinction. Nature across most of the globe has now been significantly altered by multiple human drivers, with the great majority of indicators of ecosystems and biodiversity showing rapid decline. There are obviously many conflicts built into the UN 2030 SDGs. We need more energy to reduce poverty in the developing countries. We need then renewable energy sources as biomass and hydropower to decrease the use of fossil fuels, but this will have huge impacts on biodiversity.

A major question is if we can formulate objective and necessary conditions for sustainable development which all major actors can agree upon?. Is there a non- normative necessary condition? A necessary condition is a condition that must be present for an event to occur in this case sustainability. A sufficient condition is a condition or set of conditions that will produce the event. A necessary condition must be there, but it alone does not provide sufficient cause for the occurrence of the event. Sufficient conditions will inevitably include normative conditions since sustainable development is a normative process. As examples, continuing economic growth and infrastructure development, in some subregions, are required for achieving the sustainable development goals of eradicating poverty and hunger and ensuring energy, health, and water security but needs to be pursued in harmony with nature if they are to be sustained. It is clear that the ecosystems we live on are utterly complex and difficult to model. If we affect the systems, there will be changes but we have no way to assume that there is a specific red line for our impacts. The conclusion for this is that we just have to accept that the increasing use of resources will increase environmental impacts locally and globally. The only formed principle must be to apply the precautionary principle, that is, caution practiced in the context of uncertainty, or informed prudence. As the awareness of our impacts is increasing, it is now a matter how to apply this principle in practice and which political levels are to take the lead in this process.

6. How to go from awareness to action - Think Global – Act Local

6.1. Action through international agreements

Several international agreements have been signed in order to tackle some of the urgent goals and threats for humanity and to decrease global resource exploitation. One is the Paris Agreement with focus on climate change, and the other is the UN 2030 Sustainable Development Goals. However, it becomes more and more obvious that there are serious conflicting goals in the UN 2030 Sustainable Development Goals. The most important focus in the goals is to end poverty in all forms and at the same time conserve ecosystem services. These goals should be met through economic growth (minimum 7% annual growth) with the minimum use of resources and development of renewable energy sources. It is clear that political leaders have identified the necessity to handle sustainable development in practice. This process has been going on for many years and already at the United Nations Conference on Environment and Development in Rio 1992. It was stated there that the right to development must be fulfilled so as to equitably meet the developmental and environmental needs of the present and future generations. However, there are serious limitations for going from awareness to action in practice on the highest political levels. Two years before the Rio meeting, Caldwell [9] formulated the problems which hinder political solutions for sustainable development on global level:

The prospect of worldwide cooperation to forestall a disaster seems likely where deeply entrenched economic and political interests are involved. Many contemporary values, attitudes, and institutions militate against international altruism. As widely interpreted today, human rights, economic interests, and national sovereignty would be factors in opposition. The cooperative task would require behavior that humans find most difficult: collective self-discipline in a common effort.

Sustainable development thus involves many conflicting goals where actors have different values. A conclusion from this is that sustainable development in practice has to be brought down to local levels where actors can participate in formulating the normative solutions and where support for conflicting actions can be developed. National policies have to be formulated so that support from local levels is obtained. At the same time, the local development has to ensure that global environmental restrictions can be handled in stepwise solutions.

6.2. Action through local initiatives

On the local level, urban planning is still very much characterized through sectorial planning where different experts solve problems without a more integrated view of the problems related to sustainable development. This is partly due to how we educate experts in different fields at the university. Architects and urban planners have limited knowledge of energy systems, and energy engineers have limited knowledge in urban planning. The idea is to divide the very complex issue of planning urban development by reductionist thinking.

Many of the problems with today's situation in cities are related to the way we produce and use resources, and that this industrial system has been built up during only around 150 years using cheap fossil fuels. We are addicted to the idea that we can extract unlimited amounts of resources from nature and let it end up as wastes to air, water, and landfills. The problems show up in different sectors and are a result of our lack of methods to plan cities in an integrated way. We are used to plan in a sectoral way without taking the whole picture into account.

The city is a complex system so that it resembles more of a chaotic system than a deterministic one. Because of the chaotic nature of the city, it is difficult to describe causal relations even if we have access to a lot of data. This indicates the inherent difficulties in modeling urban systems using big data, a research field developing fast in the area of smart urban development. Several papers have discussed this issue and indicate that a qualitative approach might be more fruitful [10]. A lot of work is done to model subsystems in a quantitative way, but still the problem is to go from a set of quantitative models to practical solutions in the complex urban sys- tem. We collect data but cannot use them for action plans. This was once described as "Humans may well become the first species to document in exquisite detail the factors leading to its own demise (without acting to prevent it)" [11].

The city consists of many subsystems which interplay in a complex way. This complexity has also led to the division of urban planning into several city planning departments. There is now a need for more cooperation between these in order to achieve more of cross-sectoral planning which will be discussed more in detail below. When looking at these systems, which often are managed by different units in the city, it is obvious that they are linked in many ways in an unpredictable manner. Actions in one system will produce effects in another system. It is not possible to understand in detail the whole complexity, so we still also need traditional sectoral planning methods. It is, however, necessary, when developing the comprehensive plan, to consider the cross-sector effects. This can be done by using different working methods described later on.

Tracking resource consumption in urban development inevitably at the end leads to the consumers, which are people and the source for consuming resources. It is true that industry and the public sector also consume resources, but this is all for satisfying the needs of people. Cities attract people because they offer potential for the development of their life, material, and culture. A strong force here for increasing consumption is modernity. We identify modernity with a new better life and often the old things with something we want to escape from. Changing people's behavior in order to change consumption pattern toward more resource-efficient patterns has been debated extensively in the field of sustainable consumption [12]. One conclusion is that values and behavior is interlinked in a complex way [13].

Experiences from this research show that it seems change of values precedes the change of behavior. Urban planning can here play an important role in developing urban areas which affects people's values.

To develop our understanding of resource use in urban development, we have to understand the relation between resource use and different impacts. The Driver- Pressure-State-Impact-Response (DPSIR) framework (**Figure 2**), to a large extent, intends to aid in describing environmental problems by identifying the cause-effect relationships between the environment and various anthropogenic activities in a wider socioeconomic context.

In terms of this framework, socioeconomic development and sociocultural forces function as drivers (D) of human activities that increase or mitigate pressures (P) on the environment. Environmental pressures then change the state of the environment (S) and result in impacts (I) on human health, ecosystems, and the economy. Those may lead to societal responses (R) to the corresponding drivers, pressures, state of the environment, or impacts via various mitigation, prevention, or adaptation measures with regard to the environmental problems identified. Today focus is very much on impacts and responses, but more and more we identify consumption as an important driver for environmental impacts. It is, however, easy to fall down into passivity and frustration when facing the huge environmental problems we are facing on a global level. It might be easy to say "I have not created the problems and I cannot solve them." However, today we must consider that everyone is part of the problem as well as of the solution. The expression "Think globally—Act locally" urges people to consider the health of the entire planet and to take action in their own communities and cities. Long before governments began enforcing environmental laws, individuals were coming together to protect habitats and the organisms that

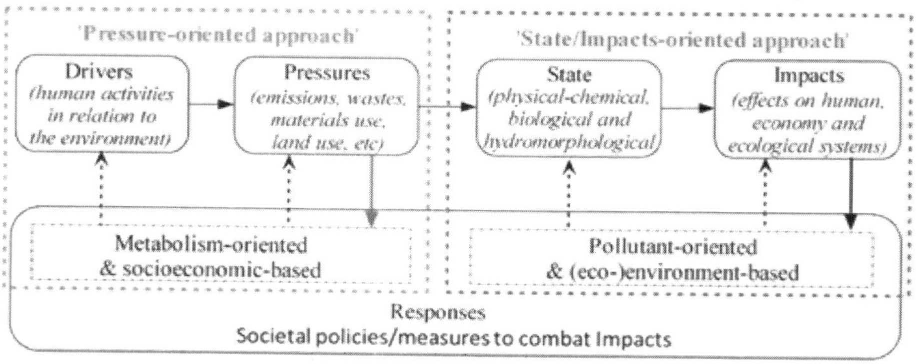

Figure 2. The DPSIR model.

live within them. These efforts are referred to as grassroots efforts. They occur on a local level and are primarily run by volunteers and helpers. We have to become more proactive in the way that we avoid creating problems instead of trying to solve them once we have created them. To do so people have to get knowledge about causal effects to develop an action-oriented approach. Many of the actions are complex in that they often move the problem. The electric car is an example of this. Mobility in a city is essential for developing a liveable city, but electricity is not a primary energy source but just a carrier of energy. If electricity is produced in coal power plants, we have just moved the emissions when introducing electric cars. Looking at the whole energy system, it might also turn out that a modern diesel car is more energy efficient than an electric car. By moving the emissions away from the city, there is also a risk that people's awareness is changed so that they believe that the problem is solved.

The development of more sustainable cities must involve actors at local level including the people who are inhabitants in the city. Human activities are the central source for resource use and impacts. This mean that we have to develop not only awareness but also new methods for integrated action plans. New methods and tools are needed for urban planning; different methods must be combined to offer explanation models for the system. The single largest limitation for this to happen is the sectoral planning system.

In the next section, we will describe such a support for planning which has been developed in Sweden by thousands of actors and which is built on gradual experience development.

7. Swedish experiences of sustainable urban planning

Sweden has a long tradition of participative urban development on local level, and several city areas have been internationally renowned [14]. In several cases comprehensive planning has started using a simple metabolic model as a platform for discussion between city planners, architects, developers, construction companies, and companies in the energy and water sector. In order to understand and visualize the urban metabolic system for a city, an eco-cycle model (as can be seen in Figure 3) can serve as a platform for visualizing cross-sectoral links between urban systems, thus stimulating discussions and creating new ideas.

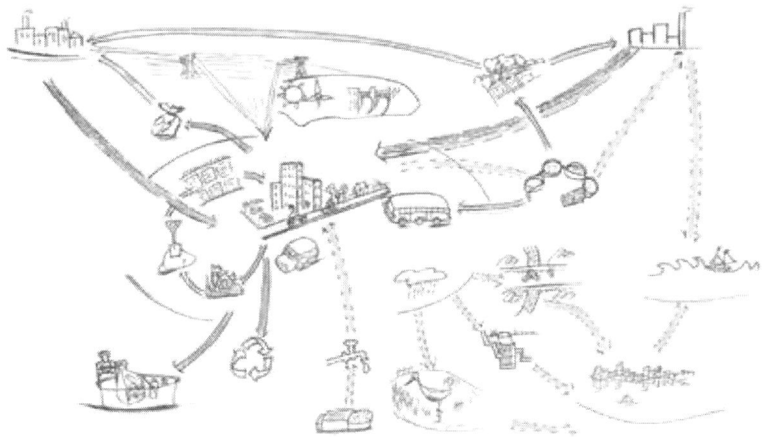

Figure 3. Example of an illustration of an eco-cycle model.

The idea behind eco-cycle planning is that what comes out of the city or city area in the form of waste (domestic or industrial waste or wastewater) should be returned to the city or city area as recycled or reused material or in the form of energy (electricity, district heating and cooling, and transport fuel). The difference between resource and waste is a matter of order (entropy) and quality. This applies both for materials and energy. In the same way as for energy, the quality of materials is gradually reduced in the city metabolism. Eco-cycles can be created at different spatial levels. Using the eco-cycle model for a systems approach permits discussions on:

- Reusing and recycling materials, e.g., construction material.
- Minimizing the use of energy from finite resources, e.g., fossil fuels.

- Using less energy through higher efficiency, passive systems, etc.
- Recovering energy through exergy ladders, downcycling of organic matter from, e.g., waste.
- Producing and storing energy through local sources, e.g., solar energy, heat pumps, and developing systems integration.

We here again emphasize the need to approach a system such as urban planning with interdisciplinary methods. Urban planning is not a system which is limited to a specific science. Urban planning is an incredibly complex system in constant change. This fact demands that different stakeholders together form an image of the system to improve their understanding of the system. It is no longer sufficient that one stake- holder alone, for example, researchers, offers solutions to problems in urban planning.

The long tradition of participative urban development on local level in Sweden has now been accumulated into a support system for sustainable urban planning called CITYLAB, which will be described below.

8. CITYLAB: a model for stakeholder cooperation in developing sustainable city areas

Based on the historical experiences from sustainable urban development, the Sweden Green Building Council (SGBC, a membership NGO) has developed

CITYLAB, a tool and certification guide for the support in the development of more sustainable city areas.

8.1. The background of CITYLAB

The Swedish industry wanted a flexible methodology for sustainable urban development that would work for big and small projects with different prerequisites. Work started in 2010, and the first part CITYLAB GUIDE for the development of sustainability programs was launched in early 2018. The development process involved thousands of people from governmental agencies, municipalities, construction companies, consultants, architects, and researchers.

Compared with other existing sustainable city evaluation and certification systems, such as LEED in the United States and BREEAM in the United Kingdom, CITYLAB pays attention to the dynamic development process of the project and supports the cooperation and experience exchange between different participants and projects. CITYLAB is a supporting tool for participative planning more than a template for certification according to standards. Early in the development of CITYLAB, four value words were established, namely that the CITYLAB certifica- tion system should *make a difference, be adaptable, be credible, and be usable*.

The CITYLAB GUIDE has an open structure that can be adapted to meet new demands and conditions. This flexibility recognizes that urban development projects take place over extended periods, often involve ambitious goals, and at times require new development measures, strategies, and solutions to be formulated in order to meet emerging local challenges. CITYLAB has specified *10 Overall Sustainability Goals* for sustainable urban development. These goals have been developed based on the UN Global Goals for Sustainable Development, the Delegation for Sustainable Cities' document "Fifteen Obstacles to Sustainable Urban Development," and Sweden's national environmental quality objectives, public health policy objectives, and Vision for Sweden 2025.

CITYLAB Action Guide specifies *17 Focus Areas* that are to guide the work in addressing sustainability within an urban development project. The Guide's Focus Areas are complex and partially overlap one another; as such, they are to be read and applied in parallel, in support of the work undertaken in relation to sustainability within a project. The 10 goals and 17 focus areas are illustrated in **Figure 4**.

8.2. Integrated metabolic thinking with CITYLAB

CITYLAB proposes 10 overall sustainability goals for urban development.

Projects can choose to use all goals or focus on some of them.

The 17 focus areas proposed in the guidelines are covering all the elements of the city which will, in a complex way, affect sustainability goals. On the left side in **Figure 4** we can see focus areas which are directly related to resources and on the right side, areas which are more related to the function of the city. In the indicator system, the CITYLAB GUIDE explicitly proposes to pay attention to the synergy and conflict between these elements and manage the coordination and conflict on the basis of fully understanding the status quo and establishing a "vision" in order to achieve the project's objectives. This completely greets the logic and thinking of metabolism in the design method. In the indicator system, CITYLAB has more specific focus points, including mixed functions, affordable housing, overall development structure, public space, social issues, circular economy, green travel, biodiversity, ecosystem services, green space elements, and low-carbon energy system. These areas have a certain integration of the previous simple elements, corresponding to the city's functional services and attributes of themselves. These comprehensive level elements indicate that CITYLAB's guidelines have transferred the metabolic theory logic in the ecological context to urban design methods and have achieved new developments in theoretical construction.

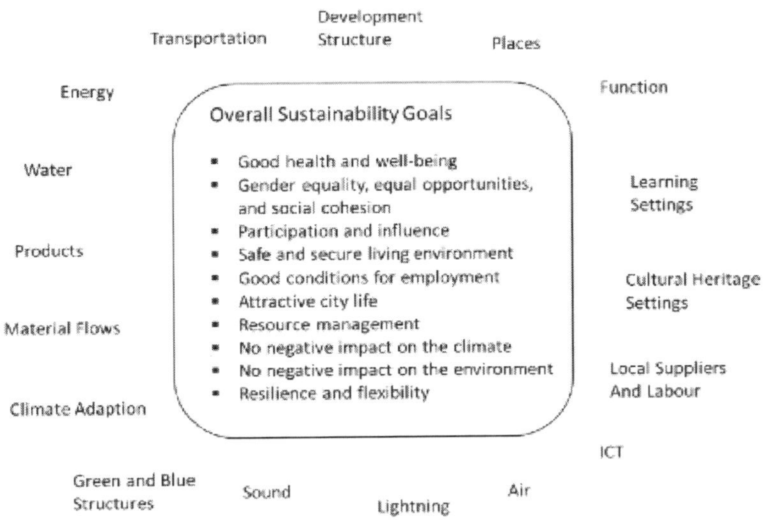

Figure 4. Ten overall sustainability goals and 17 focus areas in CITYLAB guide.

CITYLAB is a result of a long development phase in Sweden and is built on the idea of participative local planning which is typical for Swedish planning. However, the global challenges for sustainable urban planning takes place in the fast- developing countries like China where planning traditions are more hierarchical in structure. Much of eco-city planning in China is top-down starting from govern- mental policies [14]. In order to evaluate if the principles of participative processes in CITYLAB could be used in a Chinese context, a case study was developed, which is described below.

9. Application of CITYLAB in a Chinese case

In order to evaluate CITYLAB in a Chinese planning context, a research project was carried out in 2018 in Changzhou, China, in cooperation with the Swedish Green Building Council, Southeast University School of Architecture, Urbanization, and Urban Rural Planning Research Center of Jiangsu and the local government of Changzhou [15].

9.1. The objectives of the study

Changzhou is one of the most developed cities in China. Rapid economic growth and rapid urbanization have changed the scale and appearance of the city. The conflict between safeguarding ecological system services and economic development has become increasingly prominent. With an analysis of the problems of the development of eco-cities in Changzhou, the project wanted to explore sustainable urban planning methods adaptive to China and many other developing countries.

The study mainly consisted of the following aspects:

- ❖ To analyze and discuss the recent approaches and barriers for sustainable urban development in China.
- ❖ To make a comparison between sustainable urban models in China and Sweden.
- ❖ To conduct a case study in a Chinese context based on a comprehensive comparison with new sustainable urban development models in Sweden.
- ❖ To evaluate the Swedish planning model CITYLAB and to discuss in principle role models for sustainable urban planning in developing countries.

9.2. The Changzhou case

As a case of this study, Changzhou Tianning District was chosen for the following considerations:

Changzhou has a good economic foundation as a relatively developed city in China. The local government's willingness to build a sustainable city is also strong, which lays a good foundation for the sustainable development of the city.

The urban attributes and development context of Changzhou are similar to most of China's current well-developed cities. It can be regarded as a typical example, and the research results have strong reference significance for other cities. This is important for scale up of demo examples.

Changzhou not only has a long history and culture but also has unique water resources and railway resources in terms of geographical conditions. The resources that urban development can rely on are abundant, and the contra- dictions and conflicts facing sustainable development are also concentrated. Therefore, at the level of method exploration, Changzhou can be regarded as a typical example of sustainable urban planning practice in developing countries and has a strong representative significance.

The case selected by the study is located in Changzhou City, and is an important city on the extension axis of the Shanghai-Nanjing Railway, the Beijing- Hangzhou Canal, and the Shanghai-Nanjing Expressway. Changzhou has a long history and charming river city culture and is one of the most developed cities in China. The city is historically the absolute center of the Yangtze River Delta and has experienced an era of development along the canal and along the railway. It consists of six urban districts, Tianning, Zhonglou, Xinbei, Wujin, Jintan, and Liyang, with a population of 4.72 million in 2017. As a part of it, Tianning District is located in the northeast of Changzhou City, near the junction between Changzhou and Wuxi. It covers 154 km^2. The Grand Canal passes through the district which covers some historical and cultural blocks such as Qingguo tunnel and Qian Hou Bei An. There are many water networks in Tianning District. There are seven main rivers and eight branch rivers with a total length of 52.5 km. The main functions of rivers are water diversion, drainage, flood discharge, and ecological landscape.

9.3. Development modes

There are two main urban development models in China. One is the unrestricted development of satellite towns in the radiation zone of the central city which is called suburban sprawl which results in a chaotic state of the city or a machine- made urban appearance; the other is the vast towns that are far from the city and are currently experiencing the complex pattern of decline and revitalization, trans- formation, and restructuring. With the urbanization of China going deep into the water, unbalanced, uncoordinated, unattractive, and unsustainable development remains a big problem. A number of new cities almost have no features and lack vitality. Therefore, the positive impact of an independent building or micro-urban design should be given special attention which can be regarded as singularities of urban development.

The Tianning District of Changzhou City has a great potential to generate urban singularities by using its own characteristics because of the "Three Rivers and Four Cities" urban form which may be a booming district on the basis of the ecological network background and heavy breath of life. In terms of the urban catalysts, the construction of urban events should follow the principle of urban story theory; pay attention to the integration of people and nature, the integration of people and history and culture, and the integration of people's interactions, in order to pursue the overall ecological value of the city and the fair values of the environment; and finally make the city maintains long-term vitality. Secondly, the concrete strategy of "city marketing" should be put forward to shape the positive public opinion. Thirdly, a sound and scientific policy evaluation and evaluation system should be established for the influencing factors of urban events. In addition, the enthusiasm of urban residents for participating in major events and the shaping of the city image should be combined to enhance urban culture and the interaction between cities and major events. As the "urban singularity" becomes the indispensable component of urban rejuvenation, it cannot be operated in isolation and painstakingly. To achieve a healthy and sustainable development of the city, the key lies in finding comparative advantages, creating competitive advantages, and building industrial advantages [16]. In order to avoid the homogenization competition between cities and achieve differentiated development, the road of urban development and revival should pay close attention to the characteristics of the city itself.

9.4. The case study

A framework was developed as shown in **Figure 5**. The framework covers eco-city development guidelines for district and smaller plots. Forming several representative plot types, each plot type is typical of a piece of land. Next, consider the development method of a specific type of typical plot. The overall workflow is shown in **Figure 5**.

Starting from the idea of urban metabolism, the CITYLAB framework analyzes the macroscopic urban development context and the microscopic composition of urban organisms to construct a "two-way complementary" design research process. It is mainly divided into three parts: district, plot, and integration. First of all, by analysing the city's environment at the district level including social factors, government composition, economic development, etc., one can develop a general vision for the current district development. In order to further implement the general vision and provide higher practicality, based on the "bottom-up" thinking of geography, we divide the whole district by comparing the similarity and particularity of the geographical and economic patterns of the site.

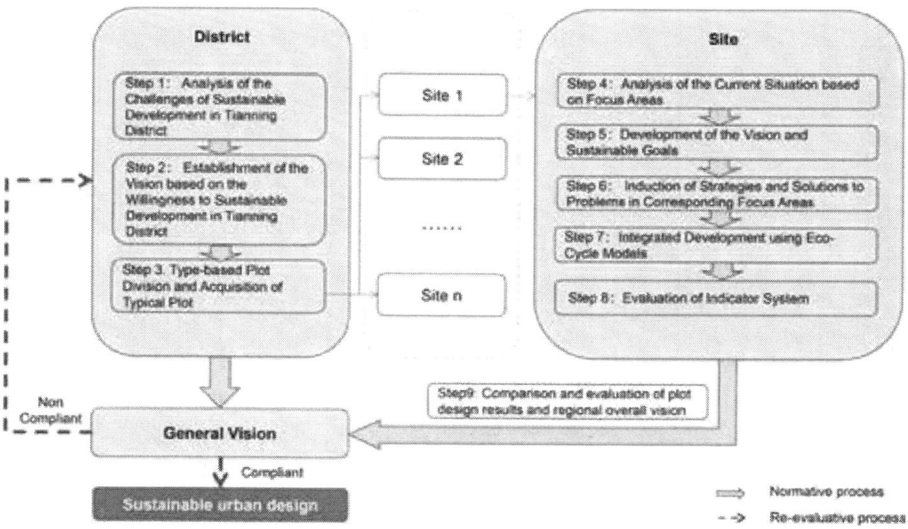

Figure 5. Sustainable urban planning workflow in Tianning District.

For the treatment of typical plots, it is more convenient and direct to deal with the city's metabolism problem, that is, by organizing the local planning process to achieve the established cross-disciplinary sustainable development goals and using these small-scale plots as the singularity of regional planning development, based on the similarity and continuity between the plots, form a series of urban renewal chemical reaction sources and activate the overall district sustainable development plan.

In this series of processes, the results of the master plan design of the plot need to be further compared and reassessed with the overall development vision of the district. This method of backtracking reverifies the overall deductive logic. If the outcome is compatible with the general vision of the district and can play a role in promoting it, then the method of analysis and derivation is effective, and the design results are in line with the needs of sustainable development. If the design result of the plot is not compatible with the general vision of the district, it is necessary to re-analyse the relevant influencing factors from the district perspective. Also, the presentation of the general vision and the division of the plot type will also be adjusted accordingly. The process covers the normative process and the reevaluative process. It has a strong correspondence and complementary relationship with China's current sustainable city development focusing on policy guidance, lack of attention to local issues, and lack of system integration.

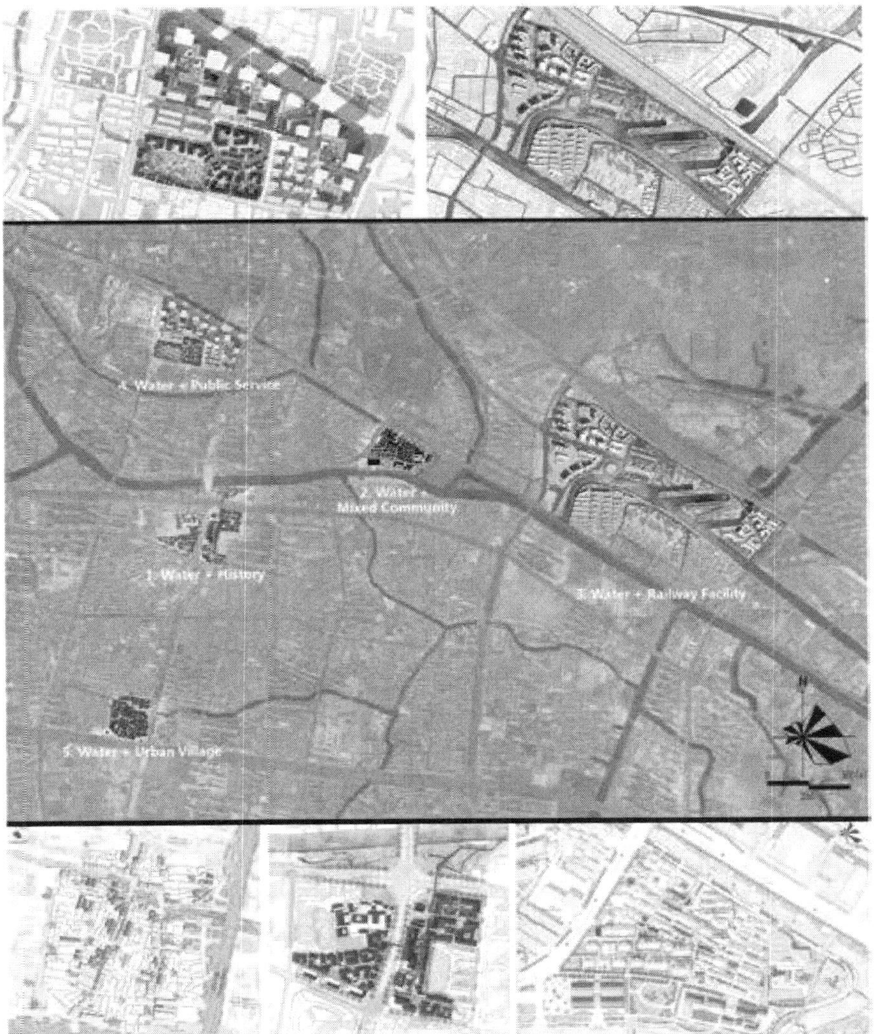

Figure 6. General plan design for sustainable planning of typical plots in Tianning District.

Based on the above method, the workshop carried out relevant sustainable urban planning on five typical plots in Tianning District, forming the results as shown in **Figure 6**. These five plots serve as demonstrations of the district, "from point to face" to activate the renewal of the surrounding plots and eventually form a comprehensive coverage of sustainable urban planning. Based on the above methods and results, we conducted an interview with the local government of Changzhou. The relevant feedback is as follows. The local government of Changzhou believes that the planning of ecological cities should first analyze the internal causes of urban ecological problems and clarify the relationship between them and the urban development mechanisms. Secondly, new sustainable development models and mechanisms should be established on the basis of the original development of the city, and they should be coordinated with the original model and at the same time promote public participation and form a long-term development model of the ecological city. CITYLAB principles can be gradually applied in the construction of Changzhou eco-city, and in the actual planning and construction. The 10 sustainable development goals should be adjusted according to the actual situation in Changzhou. The five ecological urban planning experiment of typical plots is a strong testament to the applicability of sustainable urban development planning based on the CITYLAB GUIDE.

9.5. Case study results

Based on the current situation of sustainable urban construction in China and the sustainable urban planning and design in Sweden, the main problems and challenges of sustainable urban planning and design in China are summarized. On the basis of these problems and challenges, we choose the guidelines of CITYLAB eco-city construction developed and implemented by the Sweden Green Building Council as the guidance basis and take Changzhou Tianning District as an example to carry out practical exploration, which proves the operability of the method at the practical level. Based on these guidelines, we establish a basic model for sustainable urban development planning in developing countries and lay the tone of the study.

Nevertheless, in the practice of sustainable urban planning in Changzhou Tianning District, we can realize that there are still some points that are difficult to integrate with CITYLAB and there are difficulties in the implementation process:

1. As a planning support system, how CITYLAB proposals for urban project planning integrate with the existing urban and rural planning system in Changzhou, and how to participate in the urban decision-making system?
2. In addition to cooperating with local government and local professional planning agencies, how can CITYLAB platform attract more local public participation and promote the local "bottom-up" urban mechanism in Changzhou?
3. As a system platform covering the whole process of urban planning, construc- tion, use, and maintenance, CITYLAB has a certain difficulty in intervening in the local urban and rural construction in Changzhou in a short period of time and providing corresponding technical support in planning, construction, and management.

10. Conclusions and discussions

Though 60–70% of people in developing countries still live in rural areas, half the total world population now lives in urban areas. Rural people continue to move to cities attracted by the promise of work, higher salaries, and a better social life. The fast urbanization rate together with the increasing population and consumption is challenging the long-term sustainability of our social systems and supporting ecosystems on earth. The urbanization leads to the increasing use of resources and environmental degradation on global and local levels, and many signs show that we have reached a point where the stability of the global ecosystems is in danger.

The United Nations 2030 Sustainable Development Goals are focused on reducing poverty in the world. This will require economic growth, and the big issue is if this growth can be decoupled from the increased use of virgin resources and environmental degradation.

Through globalization, resources and products are transported long distances, and it is becoming hard to distinguish between local and global effects. This increases the environmental impacts, but it also makes people feel that the overall situation is so complicated so they cannot affect it. Bringing things closer to people will create more awareness and can create enormous opportunities for new ideas and business to solve the existing problems.

The success of international agreements to handle the global problems has been limited. This is because deeply entrenched economic and political interests are involved. Political leaders are locked up to promises of economic growth and increasing welfare.

China has succeeded in reducing poverty on a massive scale through fast economic growth but to the price of environmental degradation both on local and global levels. China is trying to become a role model for sustainable urban development, but the backside of this model is obvious. China has developed several top-down strategies for sustainable urban development through the concept of low-carbon ecocity development, but with limited success.

It seems more and more urgent to develop support models for urban development on a local scale. As sustainable development involves many normative decisions, participatory planning and cross-sectoral planning will be needed to ensure that conflicts between goals can be resolved. Cross-sectoral planning means in practice bringing competences from different urban sectors together physically in workshops, in order to discuss how integration between the urban systems affects the comprehensive plan in an early stage of conceptual planning of the city.

Sweden Green Building Council has developed a recent model for the support of sustainable urban planning called CITYLAB. This has been used in a case study in the city of Changzhou in China. Participatory and cross-sectoral planning is a challenge in China where there are barriers between different city departments and between the urban planning department and state-owned companies. Planning of energy solutions when developing a plan is especially difficult since the state-owned energy companies are usually not taking part in the city planning process.

The case study in China was carried out during a very short period of time, and more experiences of integrating the new method with existing planning routines in China will be needed.

References

1. UNDP. Sustainable Urbanization Strategy; New York; 2016
2. Lidskog R, Elander I. Ecological modernization in practice? The case of sustainable development in Sweden. Journal of Environmental Policy & Planning. 2012;**14**(4):411-427
3. Meijer A, Bolívar MPR. Governing the smart city: A review of the literature on smart urban governance. International Review of Administrative Sciences. 2016;**82**:392-408
4. United Nations. World Population Prospects. ESA/P/WP/248, 2017 Revision. New York; 2017
5. Ehrlich PR, Holdren JP. Impact on population growth. Science. 1971;**171**(3977):1212-1217
6. UN. Transforming our World— The 2030 Agenda for Sustainable Development. A/RES/70/1; 2015
7. Martins NO. Analysis ecosystems, strong sustainability and the classical circular economy. Ecological Economics. 2016;**129**:32-39
8. IPBES. Global Assessment Summary for Policymakers [Online]. 2019. Available from: https://www.ipbes. net/news/ipbes-global-assessment- summary-policymakers-pdf
9. Caldwell LK. Between Two Worlds: Science, The Environmental Movement, and Policy Choice. Cambridge: Cambridge University Press; 1990
10. Kurcheeva GI, Klochkov GA. Comprehensive approach to smart urban development based on big data application. Journal of Physics: Conference Series. 2018;**1015**:042025
11. Rees W, Wackernagel M. Urban ecological footprints: Why cities cannot be sustainable-and why the are a key to sustainability. Environmental Impact Assessment Review. 1996;**16**:223-248
12. Forum WE. Engaging Tomorrow's Consumer Project Report; 2014
13. Sharma RAM. Values influencing sustainable consumption behaviour: Exploring the contextual relationship. Journal of Business Research. 2017;**76**:77-88
14. Liu H, Zhou G, Wennersten R, Frostell B. Analysis of sustainable urban development approaches in China. Habitat International. 2014;**41**:24-32
15. Zhang TEA. AQUA-URBANISM—A Citylab Guided Joint Research & Teaching Studio in Chanzhou Tianning (To be Published). Nanjing: Southeast University Press; 2019
16. Li Y. Enhancing the adaptive capacity for urban sustainability: A bottom-up approach to understanding the urban social system in China. Journal of Environmental Management. 2019;**235**:51-61

Chapter-5

Analysis of Urban Environment Sustainability in Kurdish Cities of Iran Using the Future Study Approach (Case Study: Saqqez City)

1. Introduction

There is an unequal urban growth taking place all over the world, but the rate of urbanization is very fast in developing countries, especially in Asia. In 1800 A.D., only 3% of the world's population lived in urban centers, but this figure reached 14% in 1900 and about 65% (3 billion) of people were urban inhabitants in 2016 [1]. Sustainability may seem like one more buzzword, and cities and towns like the last places to change, but the natural step for communities provides inspiring examples of communities that have made dramatic changes towards sustain- ability, and explains how others can emulate their success [2]. In the meantime, the natural step believes the root causes for unsustainability should be taken into account when designing for sustainable solutions and satisfaction of fundamental needs. These root causes are derived from a scientific understanding of our social- ecological systems—the interactions between humans in society and between humans, their organizations and the ecosystem [3]. Its biggest advantage is the concept of "simplification without reduction" to prevent getting lost in the details with the complex topic of sustainability. For almost 30 years, the approach has been implemented, proven and refined in education, research, businesses, municipalities, regional and national governments, intergovernmental organizations [4].

Since, the main theme of any geographical analysis is to emphasis on the space and its environmental concepts, any change in the relationship between humans and the urban environment can be considered as the final product of the human desire to change the settlement areas in order to overcome the environmental and spatial nature of the city. In this context, spatial changes in the urban environment can be considered as socioeconomic, environmental, political and spatial policy functions, which had different implications of the urban space according to its various conditions at different periods of time, making the concepts of space and urban environment with the fundamental transformation [5, 6]. On other hand, various aspects of human life have been affected due to the rapid development of urbanization in recent decades. So, urban sustainability is in fact related to the world sustainability. Accordingly, the role of cities and their impacts on the sustainability criteria have attracted the attention of many scientists [7]. Considering the WHO annual report, rapid, unplanned and unsustainable patterns of urban development are making developing cities focal points for many emerging urban environmental challenges. As urban populations grow, the quality of the urban environment will play an increasingly important role in public health with respect to issues ranging from solid waste disposal, provision of safe water and sanitation and injury prevention, to the interface between urban poverty, environment and health. Thus, unsus- tainable patterns of urban planning in the third world cities and urban land use planning are a driver, or the root cause, of a number of significant and interrelated urban environmental and health hazards faced by urban dwellers in developing countries. These health and environment linkages cut across a range of policy sectors and thus are often overlooked in policy making. They are a focus of this priority risks section on the urban environment [8–10].

In the view of the World Bank, it should be stated that urban environmental problems are often blamed on poverty. Most often, urban affluence is viewed as an environmental burden. Meanwhile, middle-income cities are often cited as extreme examples of urban environmental distress. A stylized account of an urban environmental transition can help explain these superficially contradictory claims in terms of the different environmental burdens typically associated with cities at different levels of affluence. However, urban poverty is not strongly associated with environmental degradation in the sense of overuse of finite natural resource bases or the generation of ecologically damaging or disrupting wastes. It is the consumption patterns of non-poor groups (especially high-income groups) and the largely urban-based production and distribution systems that serve those that are responsible for most such degradation caused by urban populations [11]. In this respect, urban environmental subjects can be joined to the sustainable urban development issues. Therefore, a successful city must balance social, economic and environmental needs: it has to respond to the pressure from all sides. A successful city should offer investors' security, infrastructure (including water, energy, population, management, environmental rules, etc.) and efficiency. It should also put the needs of its citizens at the forefront of all its planning activities. A successful city recognizes its natural assets, its citizens and its environment is built on these to ensure the best possible returns [12]. In the city context, this means that sustainable urban development is not a choice but a necessity if cities are to meet the needs of their citizens. Urban centers must be socially equitable, economically successful and environmentally sustainable if cities are indeed to be the home of humanity's future [13].

Considering the above explanations, the strategic environmental assessment (SEA) is a process to ensure the identification and assessment of environmental effects likely leading to policies, plans and programs. SEAs also ensure that environmental effects are mitigated, communicated to decision-makers, monitored and opportunities are provided for public involvement. A SEA examines the need for social progress, effective environmental protection, prudent use of natural resources and economic growth [14]. During the 1980s, ideas about urban environment-development relationships were changing dramatically and by the time of the Earth Summit in Rio (1992), it was widely accepted that environment and development must go forward together and in balance—in other words, towards a sustainable development. The Second United Nations World Conference on Human Settlements (Habitat II—the "City Summit") in Istanbul (1996) further considered this point in its global agenda for cooperation by acknowledging the direct and vital contribution that productive and sustainable cities can make to social and economic advancement. This understanding has been widely accepted among those con- cerned with the management of cities around the world, and has become the basis for new concepts and approaches to urban environment-development relationships [15]. In brief, the argument has three points:

- Sustainable cities are fundamental to social and economic development;
- Environmental degradation obstructs the development contribution of cities; and
- Environmental deterioration is not inevitable [16, 17].

Then, we can make a conclusion that urban environmental issues are multiple subjects and they are in the direct relationships to sustainable urban development and future strategies approaches. In this context, urban environment from Carson's view is an ecosystem or an environment that consists of components and elements such as resources, processes and their impacts on local plants and animal communities, human life, soil, water and air (as the natural environment), resources and processes as well as associated impacts on buildings, housing, roads and infrastructures (as the built environment), resources and processes. Besides these are also the environmental effects of human activities such as education, health and arts (as the social and economic environment) as a result of human activities and conversion of resources and raw materials to the product and requirement services in the urban scale. Therefore, urban environment was affected by these processes which can be considered as positive or negative impacts on human life in urban centers [18–20]. Accordingly, the natural step (TNS) is a naturalist approach founded in Sweden by scientist Karl-Henrik Robert (1989). In the context of urban environ- mental issues, this approach is formed from a series of strategic plans together with approaches of future studies. Therefore, future studies in connection with urban environment dimensions are some collective efforts that will be applied by analysis of the change or stability of agents focusing on the visualization of potential future and planning to achieve environmental objectives in urban spaces [21, 22]. The problems that cities are facing with today may be similar to those of the future cities but with greater compounding effects. Consequently, policy actions for developing sustainable city futures ought to be applied, tested and transferred to help solve problems for other cities [23, 24]. It can be stated that scarce researches have been conducted about the urban environmental crisis in terms of future and strategic studies. So, this study tries to analysis the opportunities and challenges of the urban environment using TNS approach (scenario and strategic planning) in the context of future study. On this basis, it is essential to pay attention to and try to solve the urban environment challenges.

2. Study area

Saqqez city covers approximately 1474.8 ha located between 46°13′ and 46°16′ eastern longitude and 36°11′ and 36°15′ northern latitude within the north-west of Kurdistan province in the northwest of Iran. According to the 2006 census, the city's population was 135,037 people, whereas its current population is about 145,000 individuals. The building area was reported to be 618.26 ha. The average elevation of the city is about 1496 m above mean sea level. Saqqez city is character- ized as a mountainous area located within Zagros Mountain range from southeast to north-west.

This area comprises about 15.5% of Kurdistan province. The difference of height between the highest elevation and the lowest point (Chehel-Cheshme Mountain 3173 m and Symone-Rood basin 1150 m above mean sea level) is about 2023 m. Saqqez River emanates from western mountains (Khan valley) and continues its path across the city towards northeast. **Figure 1** shows the location of study area in Kurdistan province, Iran [25] with a 987 km^2 common border with Iraq. The western border of Iran was specified by the border commission according to the Goldsmith Plan in September 1871 [26]. This borderline has separated parts of Kurdistan from Iran [27] and today a majority of the Kurdish population (about 25 million people) lives within Turkey (a group of Kurdish people also lives in Iran, Iraq and Syria [28]). Kurdistan province in Iran has been reported to have the lowest level of development [29–31]. Saqqez as the second largest city of Kurdistan prov- ince in Iran has faced several challenges and opportunities in the field of the urban environment. In recent years, due to the maladministration, a lot of these environ- mental capacities has been destroyed, and it is expected that in the near future, due to the widespread destruction of environmental foundations, the life continuing in the neighboring cities of this area will be faced with the serious challenges. This will be understandable once we know that Saqqez city has covered about 500,000 people by providing the needed services and at the time of environmental crisis, there will be major human and natural challenges. Also, there are many examples of these type issues in the Asian cities. Therefore, the analysis of the environmental status of Saqqez city can be a suitable model for analyzing and adopting its find- ings to the other cities in the developing countries of Asia. According to what was stated the present research has the most important for the study area.

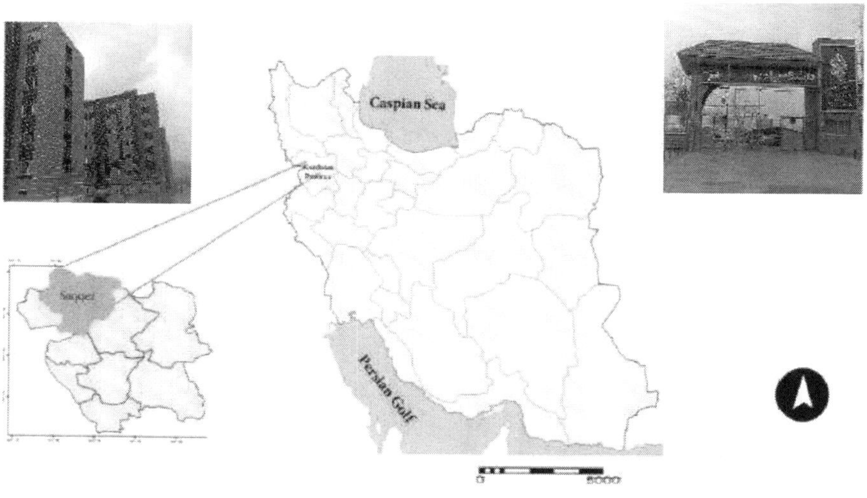

Figure 1. The study area.

3. Material and methods

The applied methodology was analytical based on its nature and practical in term of results. In addition, descriptive-analytical studies, document and question- naire in the framework of Delphi, and cross-impact matrix analysis were employed using Micmac and scenario wizard software tools. In the first step, after collection of data and identification of basic variables in the Delphi model, 50 questionnaires were distributed among 30 municipal executives and 20 academic elites with exper- tise and experience in urban management and environmental issues. Data were analyzed with a 78 × 78 matrix, by which data were included into the cross impact matrix followed by closing the CIM. After this classification, the experts were asked to evaluate the affecting variables (direct, indirect or potential) by scoring 0–3 with *p* values (0 = null, 1 = weak, 2 = average, 3 = strong, P = potential effect). Finally, a list of variables was obtained as the key driving forces, which were incorporated in the amidst cross impact method in the form of a scenario planning software. At this point, the expert can only describe the hypothesis realization probability on a scale from 1 to 5 (weak to high probability), hence avoiding any lack of precision on the expert's side. It is more helpful to see all as conditional probabilities, that is, the realization of a hypothesis in relation to other ones. Score 6 means the hypothesis is independent. Measuring the direct and indirect influence of variables was not just in the Delphi model framework, but, at the same time, to measure the various dimensions of the impact-dependence of each variable (obvious and hidden layers), it has been used for the strategically related software.

4. Results and discussion

The effective environmental indicators of Saaqez city were classified in the framework of a n × n matrix. For this purpose, 16 main categories of variables (**Table 1**) along with the 78 subcommittee variables were identified as the primary variables after holding meetings with the selected academics and executives. Then, by incorporating the variables into the mic mac software, it was attempted to define each variable according to its subsidiary and identity. In the next step, the experts were asked to score the variables based on the influence and effectiveness rate of variables on each other. As stated in the methodology section the variables were scored from 0 to 3 with P values (0 = null, 1 = weak, 2 = average, 3 = strong, P = potential effect), forming a cross impact matrix (**Figure 2**). It should be noted that due to the large size, only some parts of the matrix are expressed in here.

Based on the above findings, it can be concluded that the obtained fill rate is equal to 95.79% with two data iterations, representing a high level of variables influ- encing each other. This situation shows the efficiency of the research tool and also a very desirable verification of data collection by distributing the questionnaires.

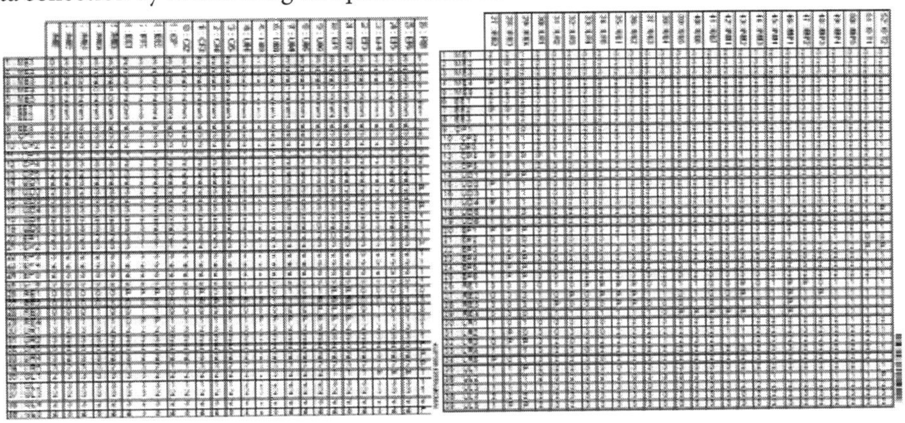

Figure 2. A small part of CIM on urban environment in the study area. Source: Research findings, 2018.

Afterwards, CIM was obtained based on **Table 1**. Since a total of 5828 values were calculated in the framework of the cross-impact matrix by the elites, 2937 cases with the highest statistical volume had a moderate impact (2) on other variables: Also, 1939 numbers had the highest impact, and 797,256 and 155 cases presented ordinary the lowest, null and potential impacts on the other variables. **Table 1** provides the categorized variables belonging to the number of abbreviated variables.

According to **Tables 1** and **2** and also by considering the analytical results, there are different practically important variables in Saqqez environmental system. **Table 3** shows various forms of the variables in the framework of determinant, result, planar and independent variables.

Figure 3 and **Tables 1–3** clarify that the religious element had the lowest influence and dependency coefficient among other variables. Also, according to calculated column rates, the management elements (ME), new technologies (NT) and recycling (RY) had the most direct dependency to other research variables.

Therefore, codification and promotion of urban recycling regulation index (RY4) with 211 scores and that of urban environmental diplomacy project (ME5) with 197 scores had the most direct impacts on the other variables. The integrated urban environmental management index (ME4) with the 188 column rates showed the highest indirect impact among other research variables (**Table 2**).

The research variables are distributed in the diagram according to the variable status in the analyses and the planar identity of some other variables. Paying attention to the findings presented in **Figures 4** and **5** can be illustrated as below.

Therefore, the instability situation of Saqqez urban environment system was confirmed at a high level through the obtained data from the analysis of indirect influences of the variables on each other, suggesting that the variables are more distributed around diagonal lines in the northeast and southeast parts of the diagram. Thus, most variables of this section have a planar identity, which, at the same time, are very influential and impressionable resulting from exacerbating or damping effects of variations due to their instable nature. Accordingly, the variables with the highest indirect impacts on the other variables include RY5 (attention to recycling economic value on the environmental management), UL7 (investment for the development of environmental land uses), RY3 (recycling of raw materials), ME5 (codification of environmental policy projects) and UL6 (biological and ecological resource management) with row points of 5,256,577, 5,194,741, 5,177,374, 5,165,320 and 5,099,651, respectively. Meanwhile, the indexes of urban integrated environmental management (ME4), compliance of Saqqez city environmental policies with the 1404 (2025) Iranian strategic planning (ME3) and the codification of environmental policy project (ME5) with column points of 5,020,651, 4,965,982 and 4,887,829 presented the highest indirect dependencies among the other variants.

Table 1. The features of primary matrix

Matrix size	Iteration	0	1	2	3	P	Total	Fill rate
		(N)	(N)	(N)	(N)	(N)		
78 × 78	2	256	797	2937	1939	155	5828	95.7924%

Source: Research findings, 2018.

Table 2. Variables of urban environment in the study area.

Categories	Abbreviation	Variables	The highest row rates	The highest column rates
1 Ecological variables	EV	EV1 & EV2	EV2: 151	EV1: 182
2 Urban recycling	RY	RY1 to RY5	RY4: 211	RY4: 185
3 New technologies	NT	NT1 & NT2	NT2: 189	NT2: 185
4 Sociocultural	CS	CS1 to CS7	CS7:155	CS7: 168
5 Management elements	ME	ME1 to ME5	ME5: 197	ME4: 188
6 Physical-spatial	PS	PS1 to PS7	PS1: 171	PS7: 162
7 Old texture	OT	OT1 to OT5	OT2: 191	OT5: 142
8 Marginalization and poverty	MP	MP1 to MP5	MP2: 183	MP4: 172
9 Population and immigration	PM	PM1 to PM4	PM4: 191	PM1: 167
10 Urban land use	UL	UL1 to UL 7	UL7: 192	UL1: 173
11 Environmental rules	LR	LR1 to LR5	LR1: 158	LR4: 181
12 Religious elements	RE	RE1 to RE4	RE4: 55	RE4: 92
13 Environmental pollutants	EP	EP1 to EP6	EP7: 173	EP4: 173
14 Urban diversity	UD	UD1 to UD6	UD6: 178	UD6: 168
15 Climate parameters	EP	CP1 to CP5	CP2: 184	CP4: 171
16 Economic elements	EE	EE1 to EE3	EE2: 188	EE1: 165

Source: Research findings, 2018.

Table 3. Different research variables according to their identity.

	Variables			
	Determinant	**Result**	**Planar**	**Independent**
1	OT	Creating urban environment	Climate parameters (CP)	Religious
		integrated plan (ME1)		element (RE)
2	RY	The impossibility of	Environment pollution	—
		establishing urban parks and	(EP)	

		green open spaces (OT2)		
3	ME	Lack of sense to place due to the immigrant citizens (OT4)	Urban land use (UL)	—
4	NT	Physical-spatial design according to UE (PS2)	Ecological variables (EV)	—
5	—	Non-standard density of housing units and population (PS3)	Urban diversity (UD)	—
6	—	Utilizing of social and cultural capacity in the UE (CS1)	Environment rules (LR)	—
7	—	Originality and environmental belongings in the metropolitan area (CS5)	Marginalization and population elements (MP)	—
8	—	Correct biological and ecological resource management (UD6)	—	—
9	—	Resource allocation and creation of new bio-capacities (EE2)	—	—

Source: Research findings, 2018.

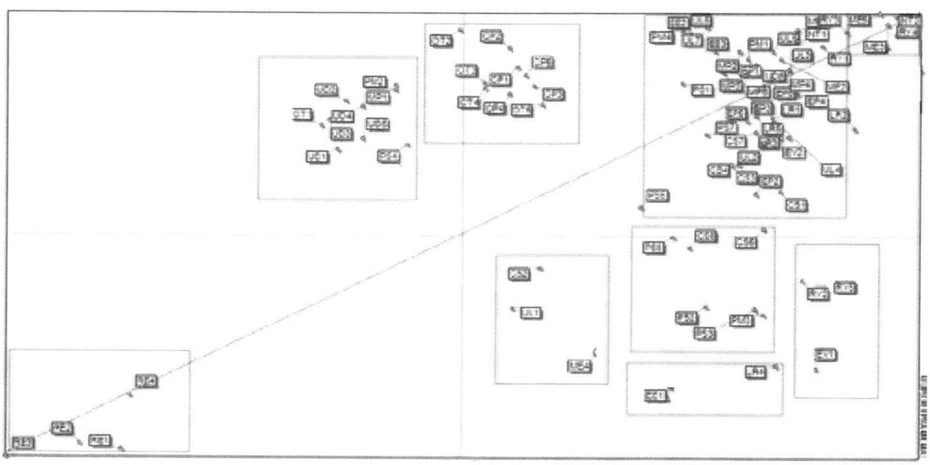

Figure 3. Distribution of variables according to direct influence-dependency and their planar identity. Source: Research findings, 2018.

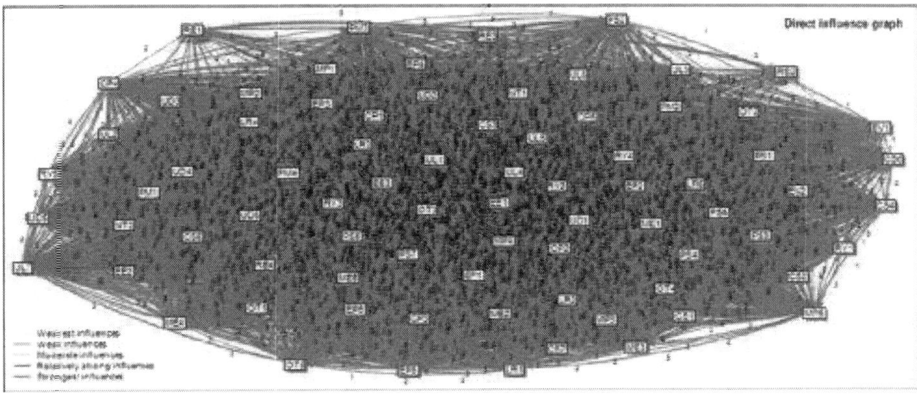

Figure 4. Distribution of variables according to their direct influence-dependency. Source: Research findings, 2018.

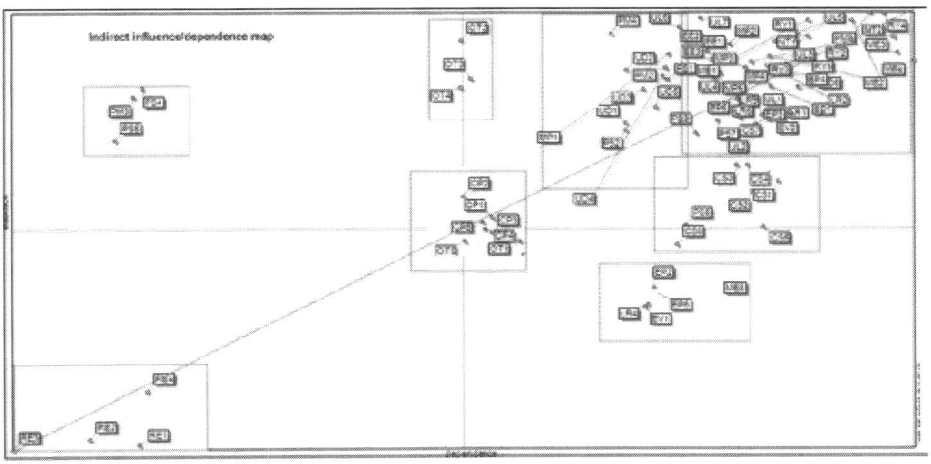

Figure 5. Distribution of variables according to indirect influence-dependency. Source: Research findings, 2018.

According to the findings of this research, it is obvious that R2 coefficient is equal to 0.0909 with a trend line value of 0.43 indicating strength relationships among the selected variables in the improvement of the urban environment in Saqqez city. Therefore, attention to the recycling of goods and waste materials with 3 weights and 0.04 coefficients besides the waste separation (Co: 0.03 & We: 4, final weight 0.12) cause lesser damages to the urban environment. Thence, the key driving forces in Saqqez urban environment are presented in **Table 4**.

By incorporating the key forces of the urban environment in the study area to the amidst cross impact methods in the frame of SMIC software, the key forces can be defined as four assumptions of consistent scenario (CS), middle scenario (MS), weal scenario (WS) and inconsistent scenario (IS). These assumptions in the form of a questionnaire were sent to 25 experts to assess their relationships by scoring from 1 to 5. Here, the expert can only describe the hypothesis realization probability on a scale from 1 to 5 (weak to high probability), hence avoiding any lack of precision from the expert's side. It is more helpful to see all as conditional probabilities, that is, the realization of a hypothesis in relation to the rest. Score 6 means that the hypothesis is independent. The software first produces a solution, namely a median probability distribution of a set of hypotheses. Minimizing its quadratic form, this is the least remote equally probable solution which would give each scenario the same probability. On the other hand, there exist certain solu- tions in Π tending to infinity, which give extreme values to each scenario. For the first eight extremes, the software calculates the uncertainty range: minimum and maximum values that can be taken by each scenario. Prob-Expert provides a series ($\pi_1, \pi_2, \ldots, \pi_r$) of r scenarios per expert, which affects the highest value of the most probable scenario.

Table 4. **The key affecting variables in Saqqez environmental management.**

Key direct influencing factors	Abbreviation	Key indirect influencing factors	Abbreviation
Urban recycling	RY	Urban recycling	RY
Management environment	ME	Urban land use	UL
Urban land use	UL	Management environment	ME
Population and immigration	PM	Economic elements	EE
Old texture	OT	Climate parameters	CP
New technologies	NT	Urban diversity	UD
Environmental pollutants	EP	Marginalization and poverty	MP
Economic elements	EE	Old texture	OT
Climate parameters	CP	Population and immigration	PM
Ecological variables	EV	Environmental pollutants	EP
Urban diversity	UD	New technologies	NT
Environmental rules	LR	Physical-spatial	PS
Marginalization and poverty	MP	Ecological variables	EV
Physical-spatial	PS	Sociocultural	CS
Sociocultural	CS	Environmental rules	LR
Religious elements	RE	Religious elements	RE

Source: Research findings, 2018.

As shown in **Table 5**, the influence and dependence between different scenarios can be obtained in the views of experts. It should be noted that the second assumption in the frame of the second scenario has the highest influence rate on another scenario and the inconsistent scenario has the lowest influence on other situations. In contrast to this case, the last scenario (IS) had the uppermost dependency to the other scenarios. According to overall approach of the research, it can be concluded that after identifica- tion of variables and assessments of their relationships by the selected executives and academic elites, a total of 78 final variables were evaluated in the framework of TNS approach key assumptions. Finally, 16 variables were selected as the key influencing and effective forces (**Table 4**). In addition, due to the proximity of some research indices to each other, these parameters were used in the context of a supplementary variable to the scenario planning process. Also, to complete the requirements of the scenario, some indices were added to the previous list and applied in the process of favorable, tragic and middle scenarios. Considering the courses of strategic planning (5–10, 10–20 and over 30 years as the short-, as medium- and the long-term periods), this research has considered the 15-year study period to explain the future study approach in Saqqez urban (**Table 6**).

Table 5. Extracted scenarios according to the probe-expert provides.

Scenarios	1 - 1	2 - Ensemble des ex
01 - 1111	0.338	0.338
02 - 1110	0.196	0.196
03 - 1101	0.072	0.072
04 - 1100	0.052	0.052
05 - 1011	0.084	0.084
06 - 1010	0.035	0.035
07 - 1001	0.023	0.023
08 - 1000	0.006	0.006
09 - 0111	0.073	0.073
10 - 0110	0.04	0.04
11 - 0101	0.015	0.015
12 - 0100	0.008	0.008
13 - 0011	0.007	0.007
14 - 0010	0.009	0.009
15 - 0001	0.008	0.008
16 - 0000	0.034	0.034

Table 6. The status of key driving forces in Saqqez urban environment.

Favorable scenario	Tragedy scenario	Middle scenario	Key element
Taking the advanced technologies and IT to enhance the urban environment development capacity in Saqqqez city	Lack the utilization of advanced technologies and information technologies (IT) to enhance the urban environment development capacity in Saqqez city	Continuation of the current situation	New technologies
Developing the recycling system and promotion of its laws by considering the recycling economic and environmental values at the highest possible level	Commercialization of environmental resources and lack of recycling system development	Intermediate stage in the commercialization and development of environmental systems	Urban recycling
Environmental development management through the codification of urban environmental rules and developing the strategies and promotion of urban environmental policy	Anti-environmental development, management through the lack of urban environmental rules	Continuation the current situation	Environmental management

The creation of sustainable land use patterns with the environmental nature	Instability of the urban environmental land use and their widespread destruction	The creation and development of urban environmental land uses with many shortcomings	Urban land use
Source: Research findings, 2018.			

5. Conclusion and suggestions

According to the presented results, the reliability of research tool was measured through holding discussion meetings, developing the key assumptions and refining the questionnaire based on the elites' views. Therefore, it can be easily understood from the findings that the urban environment system in Saqqez city is faced with such a huge volatility that continuation of the current status will lead to the forma- tion of disaster scenarios. In the best situation, if the current process continues, the urban environment will be accompanied by a widespread destruction. According to the cross matrix analysis, variables such as the management, migration, poverty, environmental and so on will be the basic resonator element of system transience in the absence of the correct uses of these variables. The research findings further signify that implementation of the foresight studies in the form of TNS approach can be an effective tool to measure the instability rate of the urban environment system in Saqqez city. According to the different parts of this research, it should be noted that solving the environmental problems of Saqqez city depends on the use of foreseeable approaches in the context of environmental models. Thus, different approaches to the achievement of sustainable urban development model in the urban environment scale will be provided using short-, medium- and long-term strategies, as well as by planning and monitoring the implementation of such strategies. To do this, the fun- damental solution to prevent environmental degradation in Saqqez metropolis is to think deeply and to have a commercialization view of the space as well as a sustainable spatial attitude towards long-term sustainable approaches. Based on the results from all extracted data in the analysis process, the value of internal factors (IFE), external factors (EFE) and weaknesses, strengths, opportunities and threat matrix, can be presented as **Figure 5**. The following propositions are presented as some safe ways:

- Preparing the urban environmental strategic plan with attention to the urban nature-oriented approaches
- Rebalancing the urban environment's capacity and the citizen uses
- Giving more roles to the citizens for improving the urban environment indicators
- Developing local and regional urban recycling centers through long-term plans
- Giving more importance to the direct and indirect impacts on the developing environmental scenarios
- Trying to achieve the optimal scenarios for effective management of the urban environment

References

1. Rahman A, Netzband M, Alka S, Javed M. An assessment of urban environmental issues using remote sensing and GIS techniques an integrated approach: A case study: Delhi, India. In: Urban Population- Environment Dynamics in the Developing World: Case Studies and Lessons Learned. Paris: International Cooperation in National Research in Demography (CICRED); 2007. pp. 181-211
2. James S, Lahti T. Natural Step for Communities. Gabriola Island, BC: New Society Publishers; 2004. pp. 203-221
3. Holmberg J, Robèrt KH, Broman G. Backcasting—A framework for strategic planning. International Journal of Sustainable Development & World Ecology. 1999;7(4):291-308
4. Hallstedt S, Ny H, Robèrt KH, Broman G. An approach to assessing sustainability integration in strategic decision systems for product development. Journal of Cleaner Production. 2010;18(8):703-712
5. Rohe WM. From local to global: One hundred years of neighborhood planning. Journal of the American Planning Association. 2009;75(2):209-230
6. Smith A. The Theory of Moral Sentiments. Penguin; 2010
7. Motlaq MA, Mesgarian H, Karimani F, Aminian AA. Sustainability assessment of urban fabrics with emphasis on the legal improvement strategies (case study: Golshan District, Mashhad Metropolis). Journal of Civil Engineering and Urbanism. 2014;4(1):59-67
8. Reducing risks, promoting healthy life. The World Health Report. World Health Organization; 2002
9. WHO. Urban population in the different countries. 2016
10. World Urbanization Prospects. United Nations, Department of Economic and Social Affairs (DESA), Population Division, Population Estimates and Projections Section, New York; 2011
11. World Bank. Poverty in the third world capitals. 2016
12. United Nations Environment Programme (UNEP). Scientific Assessment of Ozone Depletion. 2007
13. UN Stockholm Conference. The sovereign limits of global civil society: A comparison of NGO participation in UN world conferences on the environment, human rights, and women. World Politics. 1972;51(0):1-35
14. Entec UK Limited. Quantification of Emissions from Ships Associated with Ship Movements between Ports in the European Community. 2002. p. 87
15. Istanbul manifesto. Report of Istanbul manifest. 1996
16. UNEP. Monitoring Urban Transformation in the Old Cities; 2016
17. UNHS. Human Development in Urban Areas; 1997
18. Harvey KL. Ephemeral active regions. Solar Physics. 1973;32(2):389-402
19. Hay ED. An overview of epithelio- mesenchymal transformation. Cells, Tissues, Organs. 1995;154(1):56-74

20. Talen E. Do plans get implemented? A review of evaluation in planning. Journal of Planning Literature. 1996;10(3):248-259
21. Cullen G. Selection Urban Landscape. Tehran University Press: Tabibian; 1998. pp. 180-121. translation: M
22. Mozaffari M. Performance management of participatory urban development planning. In: Monthly Police Involvement. First Year, No. 5. 2009
23. Girardet H. Cities People Planet: Urban Development and Climate Change. John Wiley & Sons Incorporated; 2008
24. Phdungsilp A. Futures studies' backcasting method used for strategic sustainable city planning. Futures. 2011;43(7):707-714
25. Rahimi E. Iranian new towns and Iranian regime. SDS Journal. 2012;45:123-n 156
26. Taheri AG. History of Political Relations of Iran and England. Tehran: National Works Society; 1977. p. 952
27. Sykes P. Travel 10,000 Miles in Iran. Tehran: Loha Publishers; 1984 (Transl. by Hossein Saadat Nouri)
28. Kurdish Encyclopedia. Inside Kurdistan. Sanandaj: Royal Book; 1975. p. 97
29. Rafiee R. Site Selection for Waste Transfer Station with Regard to Urban Growth Trend (Mashad Case Study). Unpublished MSc Thesis. University of Tehran; 2007
30. Rafiee R, Mahiny AS, Khorasani N, Darvishsefat AA, Danekar A. Simulating urban growth in Mashad City, Iran through the SLEUTH model (UGM). Cities. 2009;26(1):19-26
31. Rahnama MR, Heydari A. North west border cities of Iran and regional development: A case of Kurdistan Province. Journal of Geography and Regional Planning. 2013;6(5):184-192

Chapter-6

Smart Rainwater Management: New Technologies and Innovation

1. Introduction

Water is one of the natural resources and comes as a free gift like air. However, it is a basic ingredient of any living beings. According to the Atharva Veda, a life is believed to be first evolved and nourished in water. The *Rig Veda* says water has the lifesaving medicinal quality, so it needs to be conserved. Humans cannot survive without water as it is one of their basic needs. In the body of a human being, around 70% is water. In addition to that, water is necessary on a daily basis lifestyle like drinking, bathing, washing, planting, etc. Water is also required in large quantities for different sectors like irrigation, industries, transportation, and hydroelectricity plants. All great civilizations started and prospered near water bodies.

The current generation wants smartness in every aspect of life. Hence, the concept of smart city has approached. In a smart city, the following factors are very vital such as smart grid, e-health, e-environmental monitoring, smart home, smart water quality, smart air quality, etc. Among them the more vital issue is the supply of clean and sufficient water. A substantial increase in population in cities is the main problem of their water stress because the cities are usually designed for a particular population, but in most cases the actual masses dwelling there are many times that of the designed value, whereas the resources remain the same. Therefore, all problems started with it including water too. With the explosion of human population, it is becoming very difficult for providing usable quality of water even with values. Therefore, many parts of the world are facing high to extreme water stresses. This stress is due to scarcity of underground water, mismanagement of existing water bodies, increase in river water pollution by allowing industrial and urban pollutants to flow into rivers, lack of water recycling, no clever usage, and wastage of water.

This is the common scene in most of the cities in the world. In addition to this, the water bodies are drying out. With this rate, many cities of India would come to day zero condition like Cape Town in South Africa. Therefore, there is a need of smart water management system.

Water is such a recurring type reserve that can be reused after proper treatment. However, water scarcity occurs due to mismanagement and overuse of the avail- able resource. Therefore, it is often remarked that water can be better conserved by using it judiciously.

In that manner, the irrigation sector needs revolution in water management as this sector needs the maximum share of water.

Most part of earth is covered by water bodies like oceans and seas. However, the water of these water bodies cannot be directly utilized in our daily needs like as drinking water and other purposes. Therefore, there is a persistent deficiency of water both for residential use and industrial use.

2. Rainwater harvesting

Rainwater harvesting is an expertise for collection and efficient storage of rain- water from different basement areas like rooftops of residential buildings, ground surface, rock catchments, etc. These techniques are very vast. They can be very artless techniques such as collection and storage using readily available, cheap uten- sils. They also can be some very intricate techniques such as building check dams.

These methods are mostly used for water conservation. Usually, there are two basic ways of rainwater harvesting like surface runoff harvesting and rooftop rainwater harvesting. In the first method, rainwater flowing along the surface is collected

in an underground tank. In the second method, rainwater is collected from roof catchment and stored in a tank. The harvested rainwater is the purest form of water source. So, it can be consumed directly. Rainwater collected from ground catch- ments may be poor in quality with respect to the bacteriological quality, whereas if rainwater is collected from well-maintained rooftop catchment systems and storage tanks, then that water is suitable for drinking. If water is collected from a dirty surface, then the collected water can be made utilizable by using a proper filtering system. Thereafter, it can be used for some the following purposes like drinking, culinary, bathing, laundry, toiletry purpose, watering gardens, compost making, birdbaths, recharging ponds and pools, washing vehicles, fire extinguishing, etc.

3. History

Rainwater collection technologies have a very long history. It is as early as Roman civilization. It means the period is earlier than 2000 BC. Rainwater harvesting in Asia has also been carried out since the ninth century. In rural areas of South and Southeast Asia, rainwater is collected and conserved by constructing brush dams.

In Israel, rainwater is collected and used both for farming and residential use. In Istanbul of the country Turkey, the world's largest tank is created for collection of rainwater. In India also rainwater collection and utilization are practiced since long back using conventional methods. Thailand is a pioneer in applying innovation in rainwater collection since many hundred centuries back. However, recently the most recognized utilization of the innovation is done in Africa [1].

3.1. Objectives of rainwater harvesting

i. Increase volume of water bodies

Many parts of the world have two kinds of seasons like rainy season and dry season. During dry season, there is very little or no rain. Due to this, the water bodies like pond, rivers, etc.

are dried. By using these techniques, the water bodies can be recharged, and their volume can be increased [2].

ii. Lessen flood and soil erosion

By storing rainwater, it reduces the surface runoff. This reduces the surface erosion. By capturing rainwater in reservoirs, the flood problem in large rainfalls is also diminished.

iii. Prevent overuse of underground water

As population of a locality increases, its demand for water increases too. To meet this, underground water is used. Due to this reason, the level of underground water is decreasing rapidly. By using rainwater, the demand on groundwater is reduced.

iv. Save money

Pumping up of underground water is very costly than that of rainwater harvest. So, the use of rainwater saves money.

3.2. Components

Rainwater harvesting systems consist of three basic components such as the catchment area, the collection device, and the conveyance system as shown in **Figure 1** [3–5].

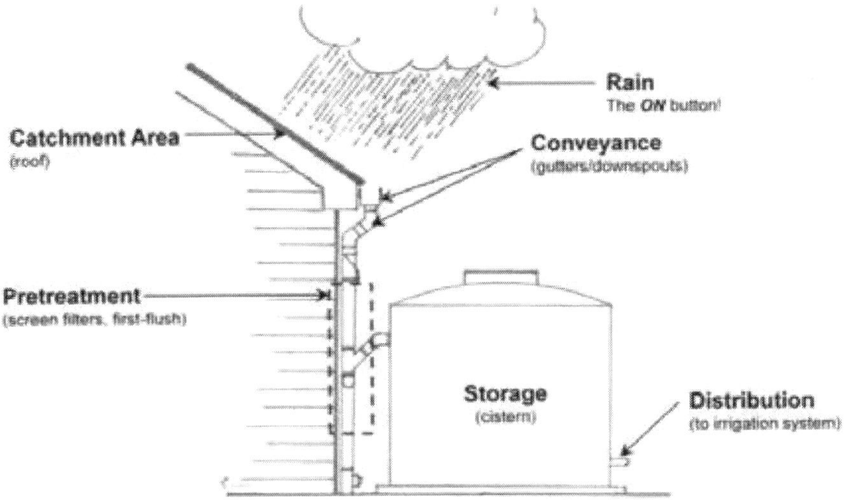

Figure 1. Components of a rainwater harvesting system (source: https://slideplayer.com/slide/9100121/).

a. Catchment area

It is the area that gets rainwater directly. From this rainwater is collected and stored. According to the types of the catchment area, it is again of two types as follows.

3.2.1. Rooftop catchment

It is the most elementary method. Here, rooftop of any building serves as the catchment. Rainwater is accumulated using easily available and cheap pots kept at the side of the roof. The quantity and feature of this collected water are influenced by the location, size, and material of the roof. A bamboo-made roof gives the lowest quality of water. So, instead of using a bamboo-made roof, it must be made up with other materials like galvanized corrugated iron, aluminum, cement, etc. The catch- ments need to be cleaned frequently to wash from dirt, leaves, and birdie stools. **Figure 2** shows a rooftop-type catchment area.

3.2.2. Land surface catchment

Here, rainwater is collected from the common surface of any ground or land (**Figure 3**). This method of water collection is also very intricate. This method can be improved by improving surface runoff capacity. That is done using a number of techniques. Runoff capacity can be enhanced by using drain pipes and storing the collected runoff water. Ground catchment area is larger than that of the rooftop area. Therefore, techniques involved with this catchment have more chance for improvement. In this method, water is kept either in small storage reservoirs or in small dams. This technique is usually applied for irrigation purpose. To increase the amount of rainwater runoff within ground catchment areas, it is required to clear or alter foliage cover, increase the slope of ground by artificial means, and reduce soil permeability by proper means [6].

The steeper the slopes of catchment areas, the quicker is the runoff and hence faster collection of rainwater. But, high-speed runoff may cause soil erosion.

Therefore, its rate needs to be controlled using plastic sheets, asphalt, or tiles along with slope. This method further reduces evaporative losses as well. Since more than 60 years ago, flat sheets of galvanized iron with timber frames have been used in the State of Victoria, Australia, to prevent soil corrosion of ground catchment area. Conservation bench terraces may also be constructed along a slope perpendicular to runoff flow for this purpose. The soil of the catchment area must be made hard and smooth. Surplus runoff water is directed to a lower collector and stored there. In addition to this, soil treatment using sodium helps in reducing soil permeability [7].

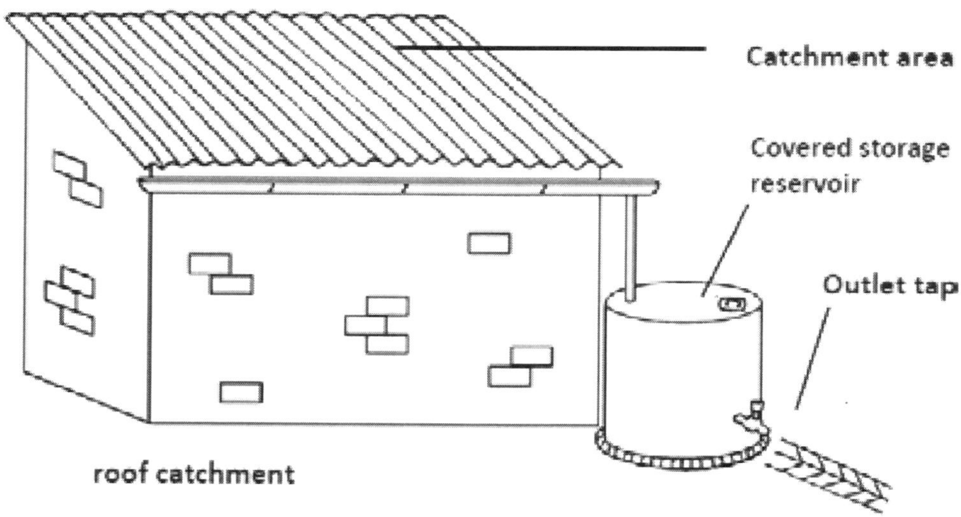

Figure 2. Rooftop catchment system (source: CTCN site).

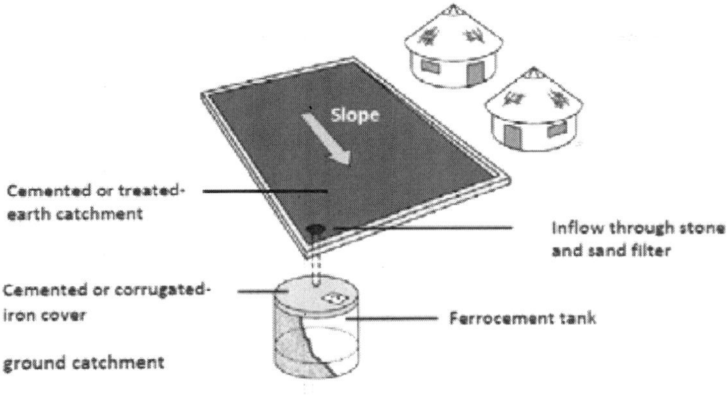

Figure 3. Ground catchment system (source: CTCN site).

b. Collection and storage equipment

The collected rainwater from the catchment area is required to be collected and stored in proper collection equipment. It can be a storage tank or a rainfall water container. Storage tanks for this purpose may be placed either above or under the ground. The tank needs to be fitted with tight cover for preventing algal growth and the breeding of mosquitos. Measure must be taken to reduce contamination of the stored water. Storage tanks can be cylinder-shaped containers made up of ferro-cement or mortar. The former container has a slightly armor-plated concrete base. On this base the cylindrical tank is mounted. Again two layers of light wire are enfolded over this tank surface as shown in **Figure 4**. This serves as a frame for the container. The latter type container is a large vessel made up of mortar. This vessel is also wounded with light wires (**Figure 5**). In some cases, the storage tanks are concrete tanks (**Figure 6**) or plastic jars (**Figure 7**). The storing capacity of rainwa- ter is calculated considering different factors like the dry spell span, the volume of rainfall, and the consuming demand [8, 9].

Figure 4. Galvanized iron tank. (source: https://www.rainharvest.com/rainflo-3400-gallon-corrugated-steel-tank- rainwater-harvesting-package.asp).

c. Conveyance structures

These structures are the means to transfer the collected water from catchment surface to the storing vessels. This structure consists of a number of downpipes attached to the rooftop gutters (**Figure 8**). Water collected from the first rain may consist of dirt and debris. The conveyance structure is required to do the primary treatment to the collected rainwater for clearing those impurities. In one of the conveyance structures, water of first rain is directed to the storage vessels after carrying out the primary treatment in these pipes. This type of structure can check the quality of the water collection. It also has a provision of manual operation of the flap.

In another method, the operation of the flap is automatic. Here, a funnel-shaped device is assimilated within the downpipe structure at a place. A gap is intentionally kept between the funnel structure and inner side-wall of the downpipe. Therefore, rainwater passing through funnel gets filtered, whereas the excess of rainwater is ejected out to the ground through the gaps. At the start of rain, a small quantity of rain passes to the storage tank due to the collection of dirt. After some time, this collected dirt is flushed away to the ground through the gaps. Then, funnel can send more volume of water through pipes to storage vessels as seen in **Figure 9**. Instead of metals, this system uses plastic, PVC, or other inert substance pipes to protect from acidic water [10, 11].

Figure 5. Ferro-cement tanks.

Figure 6. Concrete tank (Source: Taanka Wikipedia). (source: https://www.rainharvest.com/rainflo-3400-gallon- corrugated-steel-tank-rainwater-harvesting-package.asp).

Figure 7. Plastic tanks (Source: Taanka Wikipedia).

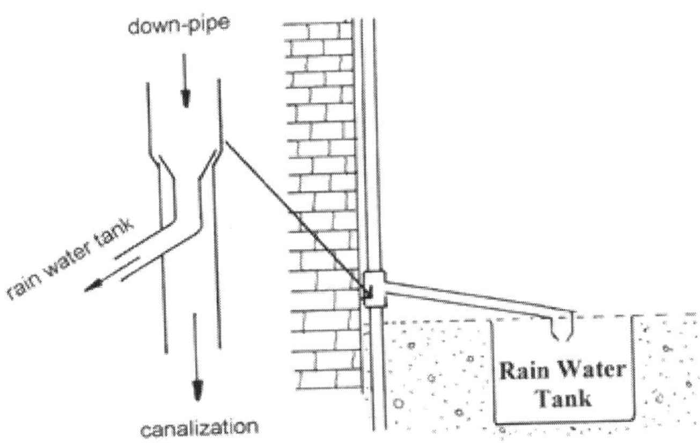

Figure 8. Conveyance system (source: Newsletter and Technical Publications, UNEP).

d. Filtering system [12, 13]

To make the collected rainwater usable, it must be contamination free, safe, and inexpensive. For that a properly constructed water filter must be used. The following filtering system may be used such as sand gravel filter, charcoal filter, and PVC—pipe filter and sponge filter.

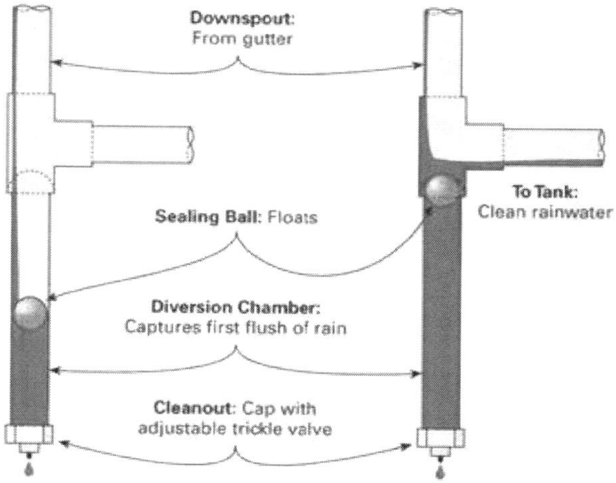

Figure 9. First flush diversion (source: https://slideplayer.com/slide/9100121/).

Sand filters are usually the most widely used filters. These are brick masonry structure. Small stones, gravels smaller than stones, and sand are filled in this structure to serve as filtering medium. Each of these filtering media is made of one filter layer. Each one of these layers is divided using wire mesh. It is shown in **Figure 10**.

Charcoal filters are the drum- or chamber-shaped structure. These are filled up with charcoal in addition to small stones, even smaller gravels, and sand layer by layer. Here also layers are divided using wire mesh. It is shown in **Figure 11**.

PVC filters are made up of PVC pipes filled with sand and gravel separated by a layer of gravel (**Figure 12**).

Sponge filters are the most simple and inexpensive rainwater filter. It consists of a PVC drum with a sponge layer in its midway. It is appropriate for residential usage (**Figure 13**).

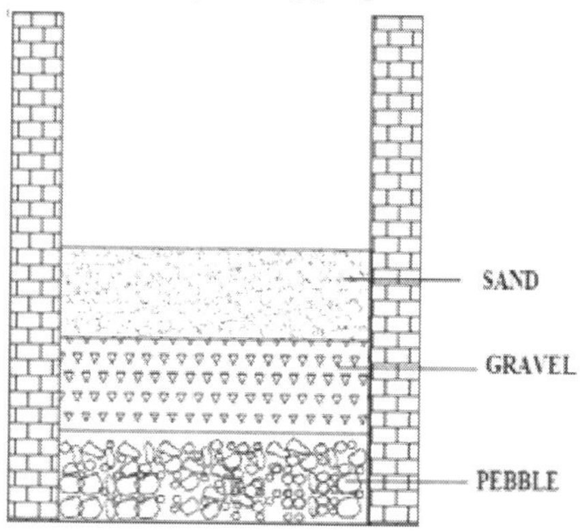

Figure 10. Sand filter.

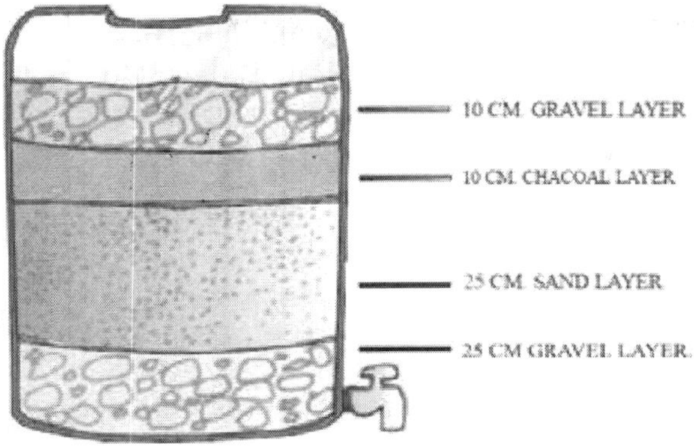

Figure 11. Charcoal filter.

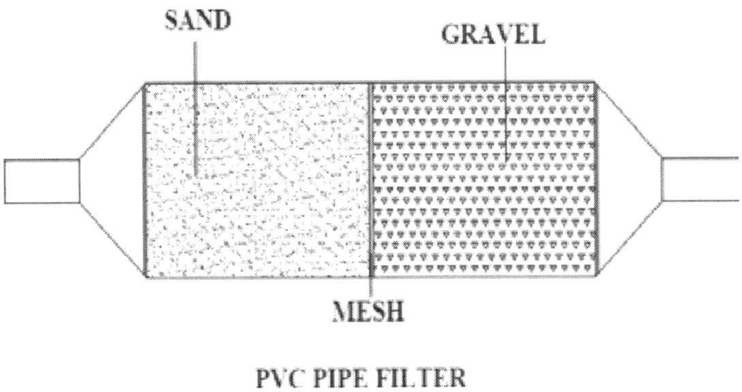

Figure 12. PVC pipe filter.

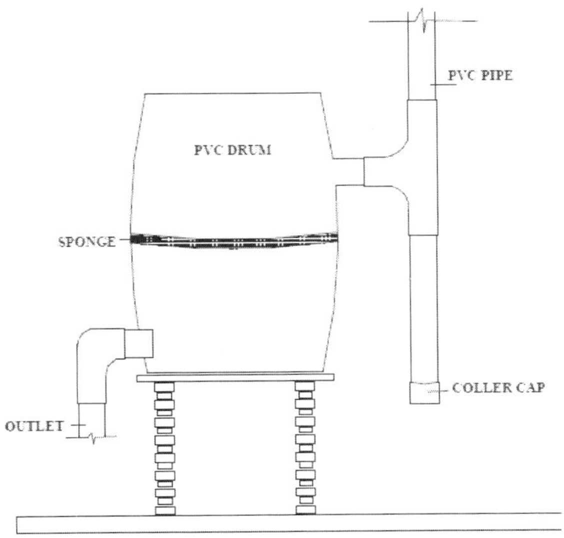

Figure 13. Sponge filter (source: Rooftop RWH by M. Patil).

4. Methods of rainwater harvesting

1. Domestic rainwater harvesting

In this kind of harvesting, the catchment area is mainly rooftop area (**Figure 14**). Here, it is required to check two factors such as rainwater storage capacity and harvesting capacity. Harvesting capacity can be improved by using the right kind of catchment area. The storage capacity can be enhanced by considering appropriate storage vessels. Nowadays, commercial domestic rainwater harvesting systems are available. These systems are categorized according to the design and capacity of their storage vessels. The designs of these systems are very trustworthy to be used. These water systems are usually used as secondary water source in addition to the main water supply. This harvesting system is generally used for watering kitchen gardens. This water can also be used for other residential usage like bathing, cleaning, washing, etc. This system is simple to install and easy to manage [14–16].

2. Rock and other surface catchment systems

Catchment areas made up of rock are the ideal type catchment systems for rainwater. These structures are constructed at a site where users have easy access. The sites should not be a place where there is a possibility of soil erosion and negative impact to vegetation. In this case, the collected rainwater is usually stored in barrages or large open containers. These are suitable for use in school, industries, colony with a large number of residential buildings, etc. In this case water treatment is required before using it. In this case, underground tanks are also applied [17] (Figure 15).

3. Subsurface dam and sand dam

In sand dams, runoff rainwater is stored in various types of small-scale reservoirs [18]. The reservoir structures are like mud, stone, or concrete dams or a simple

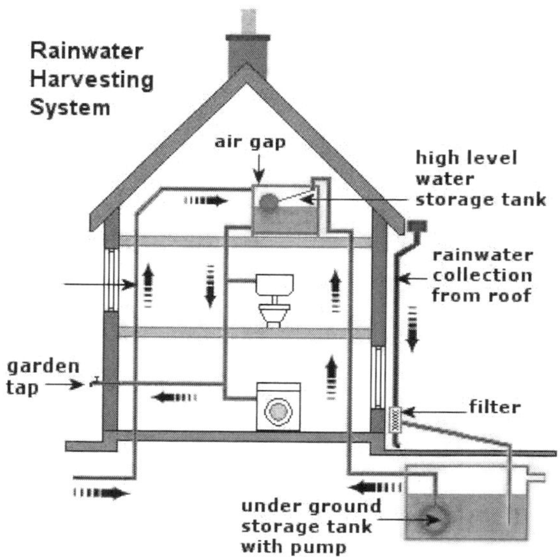

Figure 14. Domestic rainwater harvesting (source: David English Water Pumps).

Rock catchment structure at Musul, Laikipia district, Kenya (from 2001)

a

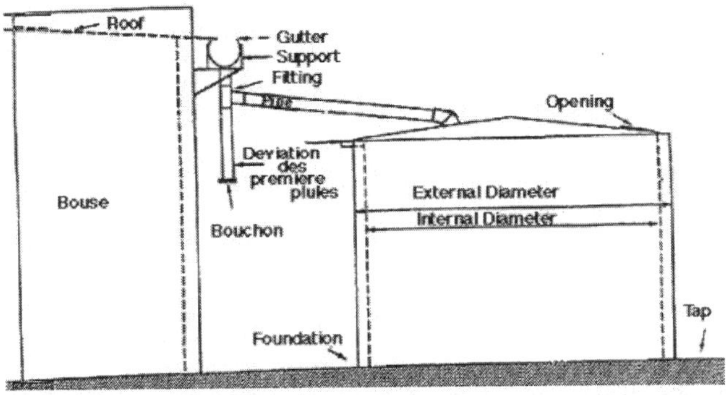

b

Figure 15. (a) Picture of a rock catchment system and (b) layout of other surface catchment systems (source: UNEP and Rainwater Harvesting Handbook).

mined pond. Another form of such structure uses hand-dug wells and some hori- zontal inflow pipes for water mining. These are called subsurface dams (**Figure 16**). Here, water comes with its purest form and hence does not need further treatment before consuming. These types of structures are very effective functionally in arid and semiarid areas where rainfall varies from 200 to 750 mm. The harvesting water is mainly utilized for domestic, cattle, and kitchen gardening. Picture of a sand dam is shown in **Figure 17**.

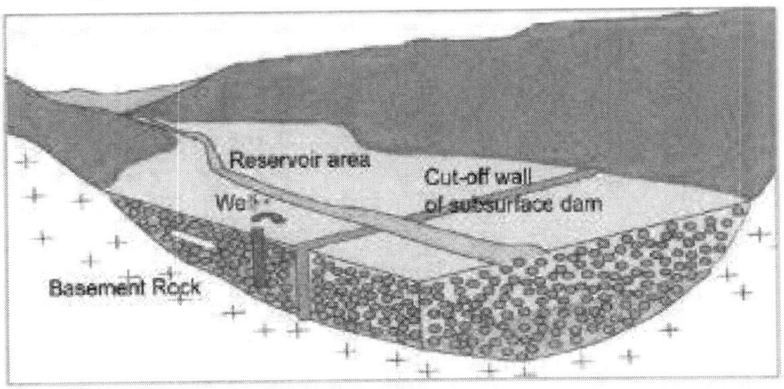

Conceptual Diagram of the Subsurface dam

Figure 16. Layout of a subsurface dam (source: samsamwater.com).

Figure 17. Picture of a sand dam (source: rainfoundation.org).

4. Earth dams like ponds and pans

Earth or mud dams are capable of storing rainwater up to 10,000 m^3 [18]. The depth of these ponds is around 5 m. These are generally built using human power or using the help of domestic animals. In some cases, farm tractor, crawler, or bulldozer are also used. Such dams are categorized into three types such as Charco dams, Hillside dams, and Valley dams. Charco dams are constructed in an almost flat surface. Hillside dams are designed for sloppy hilly surface. Valley dams are constructed in a valley surface where water is available only seasonally. Ponds are the man-made-type small earth dams as seen in **Figure 18**. These are constructed by digging the earth surface and placing that mined soil on the lower side of the pond to increase the storage volume. Pans are nature-made ground dip curves where rainwater can be gathered in rainy seasons. This is the earliest form of rainwater harvesting system created by nature (**Figure 19**).

5. Recharge structures

Surface water has a natural tendency to recharge into the underground water reserve [19]. Nowadays, this recharging is done using artificial recharge struc- tures utilizing suitable civil construction techniques. It augments cover-up of the depleted aquifer. It helps in water conservation for future usages. This recharging should not dilute the quality of the existing groundwater. The main purpose of artificial recharge of groundwater is to restore the quantity of the underground water due to excessive use.

Figure 18. Picture of a pond.

Figure 19. Picture of a pan (source: jkuat.ac.in).

Commonly used recharging methods are as follows:

1. Recharging bore wells
2. Recharge pits
3. Soak away or recharge shafts
4. Recharging dug well
5. Recharge trench
6. Percolation tank.
7. Some of these structures are shown in below figures (**Figure 20**).
8. Conservation tillage

This method is one conventional form of farming [19]. In this method, farm surface is plowed with tractor or bullock plow to loosen soil before cultivation. If certain modification is made to the conventional form of tillage as shown in the following figure, then it becomes conservation tillage. This type of modern tillage is capable of conservation of water, soil, and other energy resources by decreasing the tillage intensity and preservation of crop residue. This tillage rarely disturbs the soil surface used for harvesting of crops. A comparative view of both the tillage is shown in Figure 21.

1. Planting pits

Planting pits is also an innovative rainfall harvesting method (**Figures 22** and **23**) [19]. It holds rainwater in its pits and stops rainwater from runoff. This helps in enhancing infiltration of groundwater into underground. It also helps in decreasing soil erosion. Here, the distance between two nearby pits is kept at 50–100 cm. The depth of each pit is usually kept between 5 and 15 cm. This technique is suitable for soil with low permeability like clay. This method can be applied to semiarid areas for crops like sorghum, maize, sweet potato, bananas, etc. The major main benefits of this method are that it is easy to construct, use, and take care. However, large effort is required for construction of pits as they cannot be mechanized.

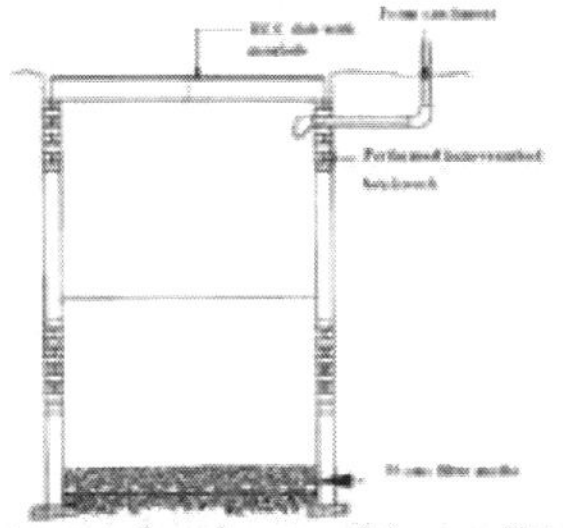

Natural recharge of dug well (Source: Ministry of Water Resources)

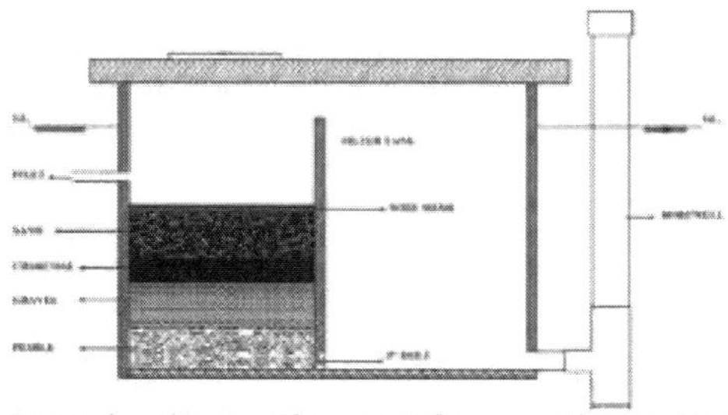

Natural recharge of bore well (Source: Ideas to Make A Rain Harvesting System In India)

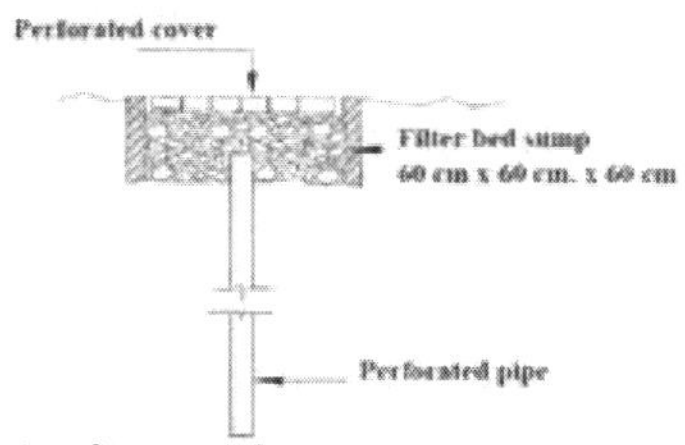

Recharge Pits

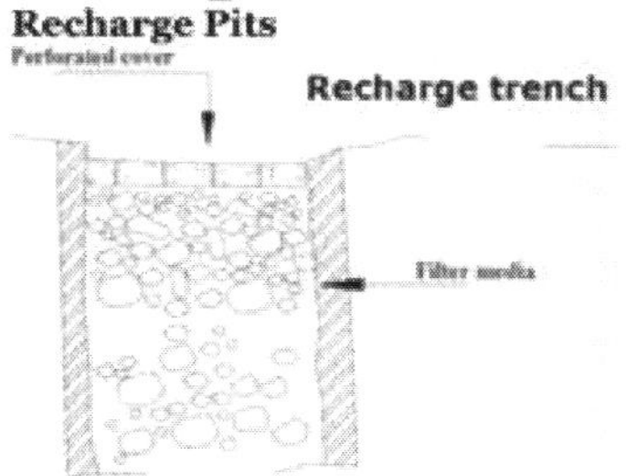

Recharge trench

Figure 20. Some of the recharging structures (source: Rooftop RWH by M. Patil).

8. Katumani pit

Katumani pits are small crescent-shaped pits side covered by native grasses and legumes [18, 19]. Each pit is 15 cm deep and 20 cm wide with downslope ridges of about 30 cm height as shown in **Figure 24**. Pitting can be extended down the slope as per the requirement and convenience. Although it is a laborious method, it conserves soil and can meet water demand of high water demanding crops like banana and sugar canes.

9. Semicircular bunds

This is one of the micro-catchment types of rainwater harvesting system [18, 19]. These consist of semicircle-shaped stone banks. The landfills of the bunds are prepared in such stagewise orientation manner in rows (**Figure 25**). If ever rainwater overflow occurs, then it will flow from one row to the next down row. Labor cost, slope of land, availability, and size of stone are some major factors of this technique.

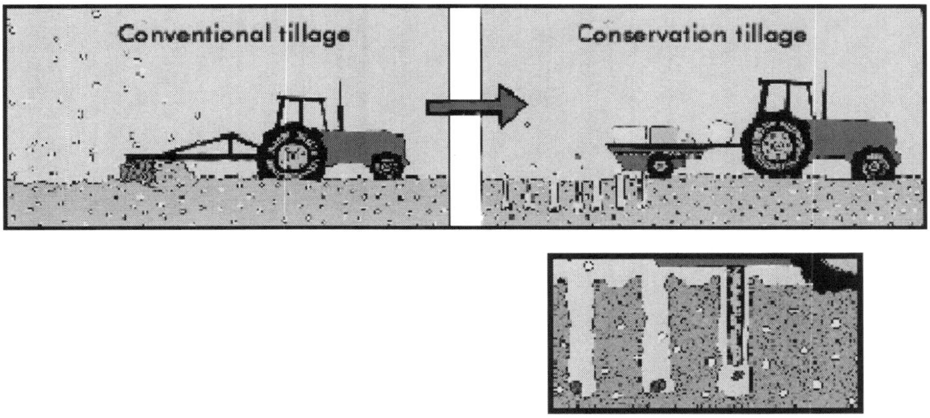

Figure 21. Comparison of conventional tillage and conservation tillage (source: Climate Tech Wiki).

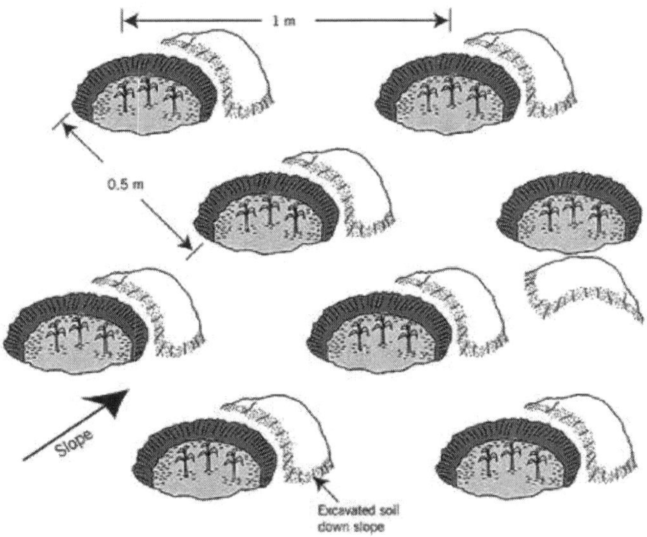

Figure 22. Planting pits (Source: Malesu et al. [20]).

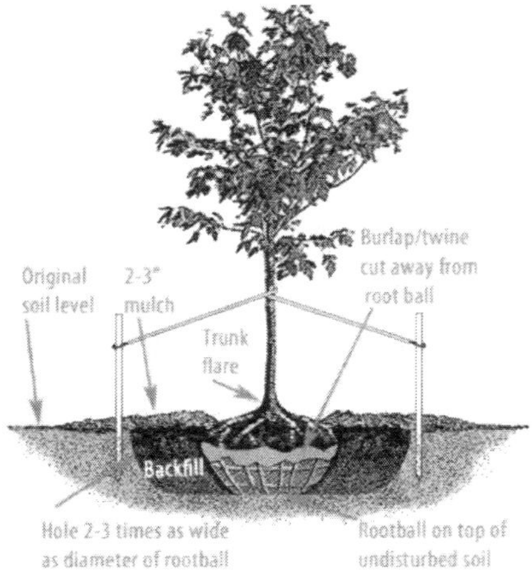

Figure 23. Digging hole for planting pit (source: https://www.gardeners.com/how-to/tree-planting/8741.html).

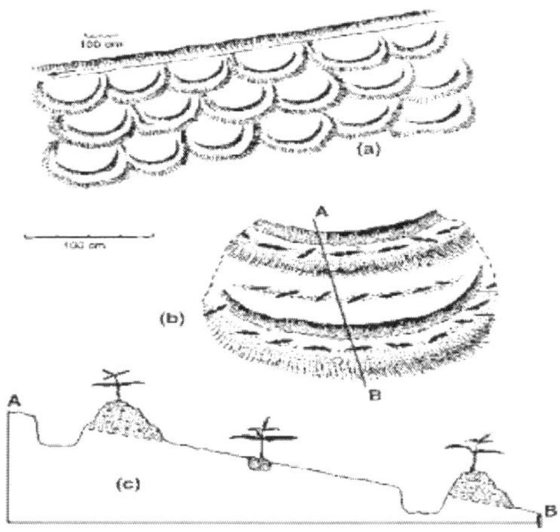

Figure 24. Katumani pits in three plans such as a, b, and c (source: Namirembe et al. [21]).

Figure 25. Semicircular bunds for fruit plant (source: Snobar et al. [23]).

10. Negarim micro-catchments

The word "Negarim" is derived from the Hebrew word "Neger" that means "runoff." Negarim micro-catchment is one of the most recent innovations of rainwater harvesting systems [18, 19, 22]. These look like diamond-shaped bowls (**Figure 26**). The boundary surfaces of these structures are bordered by small earth bunds. Each structure has a groundwater penetration pit in its lowermost corner. Runoff rainwater within each structure is accumulated and stored in their respective pits. These techniques are basically used for growing plants or scrubs. This technique is suitable for small-scale vegetation in dry areas. It helps in soil conservation too. These are easy to construct and structurally are well-ordered and accurate. The first ever structure of such micro-catchment is reported from southern Tunisia. However, it is widely used in the Negev desert of Israel.

11. Tied contour ridges

Contour ridges are small-sized clay ridges with height of around 20 cm [18, 19]. It has an upslope channel that helps in flowing runoff water from catchment strip between the ridges (**Figure 27**). Cultivation is not done in the catchment strip.

Therefore, the purpose of this structure more satisfies controlling soil erosion rather than rainwater collection. The main purpose of this system is to collect local area runoff rainwater and then keep them in land surface near to the plant roots. These structures are not meant for controlling overflow. Hence, the system is needed to be designed with a cutoff drain. This technique is practiced widely in African countries like Kenya, Niger, and Zimbabwe. Its maintenance cost is less if it is well designed.

Figure 26. Negarim micro-catchment structure with plantation (source: Development-durable.org).

Figure 27. Tied contour ridges (source: Muslim Science).

12. Contour stone bunds

These are constructed in sloppy area where rainfall varies between 200 and 750 mm [18, 19]. These structures are made of stones (**Figure 28**). So, they are suitable on stony land. These structures can slow down runoff and protect soil from filtering out. They also help in enhancing the infiltration of runoff water. These structures can be constructed both with and without spillways. Structures without spillways are simple in construction and easy to maintain. These types of structures are suitable for small-scale irrigation in farming lands.

13. Fanya juu terraces

In Kiswahili language, in Swahili, the word "juu" means "up" and the word "Fanya juu" means "throw the soil up." Fanya juu terraces are constructed by digging soil and pitching it upslope, making a mound for water collection [18]. The mound creates a runoff barrier, whereas the trench holds the runoff rainwater. The ridges are usually alleviated with fodder grasses. In these techniques crops like banana, citrus, and guava can be cultivated. This technique is suitable for areas getting low annual rainfall areas and with less than 20% slopes, hilly area, and deep soils with massive fear of soil erosion. Farmers basically use this technique to increase their crop. These terraces are perfect for cultivation of fodder grasses. In addition to it, these structures support in preventing soil erosion. Cultivation becomes easier as these terraces are made of multilevel structures. Again, yield from the crops enhanced when fertilizer is applied to it. The structure is shown in **Figure 29**.

Figure 28. Contour stone bunds in Kenya (source: Yazar et al. [24]).

Figure 29. Bean being grown on Fanya juu terraces in Lanzi (source: Michael Rastall, weadapt.org).

14. Fanya chini

The structure of Fanya chini is the same as Fanya juu. However unlike Fanya juu, here soil is put on the lower side of the contour trench [19]. These are used to conserve soil as well as divert rainwater. Here, the ridges are used to grow feedstuff. These are easier in development than that of Fanya juu tarraces. But here in this case multilevel structure is not possible. The maximum limit of slope in these structures is around 35%.

15. Earth bunds with external catchment

Bunds or teras are some small barriers made to prevent runoff water from entering a catchment with crops (**Figure 30**) [18]. Bunds reduce speed of runoff rainwater on and increase recharging of the groundwater. Several different struc- tures of bunds are found like rectangular type, contour type, and base type. In rectangular bund structure, three sides are bounded, and the fourth one is open to collect water. In contour-type bunds, row structure is erected along a hillside. The size of base bund structure is with length of 50–300 m and width of 20–100 m.

Here, bunds are constructed not to collect water. Their main purpose is to bal- ance the hydration of soil and also recharging under groundwater. The structure without a spillway may break the sides of the bunds. Therefore, spillways are also added to these structures. This helps in improving the efficiency and lowering the maintenance costs.

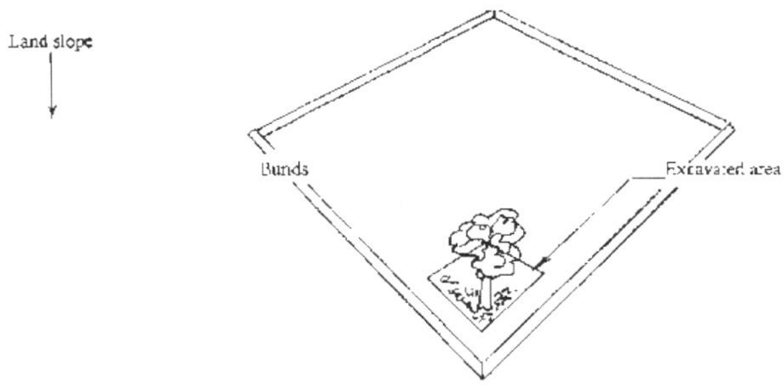

Figure 30. Earth bunds with external catchment (source: nzdl.org).

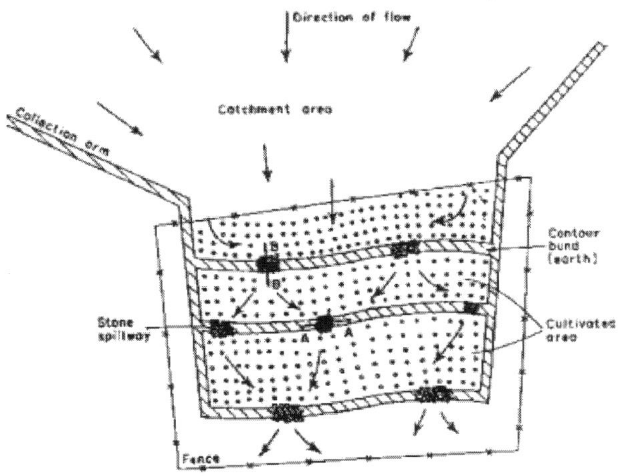

Figure 31. Macro-catchment water harvesting system (source: Imbira [25]).

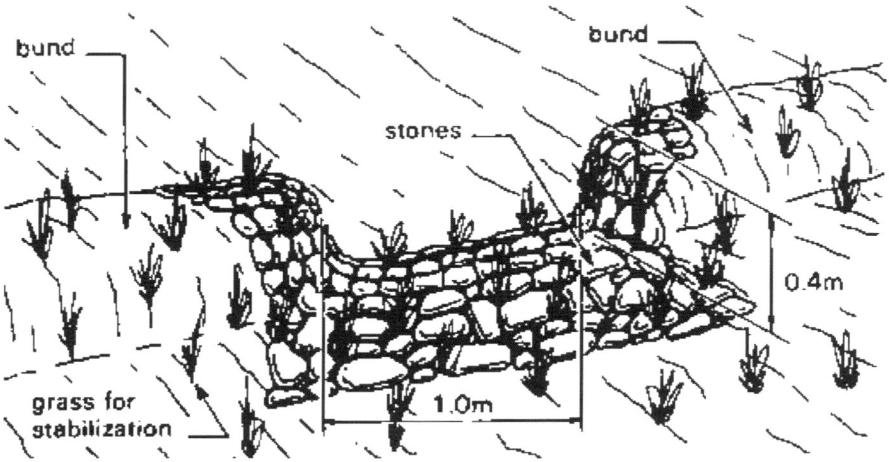

Figure 32. Spillway construction for bunds with external catchments (Pacey and Cullis, 1991).

16. Contour ridges with external catchment

This technology is the same contour ridging technique but uses an external catchment (**Figure 31**) [17]. It includes an extra spillway structure that can handle the excess runoff water flow (**Figure 32**). These structures are constructed with mud or pebbles.

5. Traditional methods of rainwater harvesting in India

Indian administrations are taking many new projects to locate and to revive the age-old conventional water harvesting methods [26, 27]. These approaches are found to be very easy, reliable, and eco-friendly. The existence of many such ancient structures has been proven by archaeological findings. In past found evidences, it is verified that the Indus Valley civilization is pioneer in it. Indians have built many more such innovative and effective structures of rainwater harvesting. Although these techniques are not very popular today, some of them are still in use. Some of such structures are discussed as follows.

Jhalaras are the stepwells having stairs in its sides. Shapes of these structures are almost rectangular. These structures collect the buried discharge of a nearby water reservoir, pond, or lake.

Talabs are either pond or pans. They are utilized for meeting consumable water for drinking and domestic work purposes. These structures can be natural or man-made.

Bawaris are a type of unique structure of stepwells. These were usually used in the ancient cities of Rajasthan.

Taanka is a traditional structure of rainwater harvesting seen in the Thar Desert region of Rajasthan. This structure consists of a cylinder-shape underground well. It collects rainwater from rooftops, courtyards, or artificially prepared catchment flows and sore for future use.

Ahar pynes are found in South Bihar. These are artificial lakes in which three of their sides are ridged, whereas the fourth side is open. The open side is joined to the end of a diversion channel as in pynes.

Johads are the small-size muddy check dams that are built to store rainwater.

They also help in recharging groundwater.

Kuruma tribe (a native tribe of Wayanad) uses a distinct structure of well called the **panam keni**. This structure is built using wood.

Khadins are also very inventive structures built in hilly areas. Water of these structures is used for cultivation purpose.

Kund is a structure having a plate-shaped catchment area. Its surface is gently inclined toward the center where a well is dug. The well is circular shaped. It is usu- ally used for supplying drinking water.

Baoli was usually built for philanthropic reasons. From baoli, everyone could draw water.

Some more such structures are **Nadi, Bhandara Phad, Zing, Kuhls, Zabo, Bamboo Drip Irrigation, Jackwells, Ramtek Model, Pat System, Eri**, etc.

6. Rain centre

The Rain Centre is a chain of organization mainly to spread literacy on rainwater harvesting and utilization. Nowadays, it is becoming very popular among urban Indians. It educates the significance of rain in Indian life. It also tells about how rain influences the customs, traditions, economy, and politics of different parts of India. This center also cites the various distinct examples of harvesting rainwater and people working in this field. The rain centers are mainly built to educate people to harvest rainwater in a proper manner. For this purpose, the local NGOs and dedi- cated citizens of an area are selected and first educated. The first rain center in India is established in Chennai in 2002.

Now, it has a chain of 11 model rainwater harvesting projects in Delhi. The Rain Centre builds rainwater harvesting systems for individual or organization with their demands. The fourth rain center is established in Burdwan, West Bengal of India.

Now, offices of rain centers are operating in West Bengal, Gujarat, Chennai, and Meerut of India [5].

7. Advantages

As per the previous all discussion, rainwater harvesting is very beneficial. Some of the remarkable advantages in implementation of this technique are marked as follows. Rainwater harvesting technologies are simple to install and operate. Highly skilled manpower is not required for implementation of this technology. Local people can be easily trained to construct and maintain rainwater harvesting plants. Construction materials are also usually locally available. It is one of the most convenient system as it provides water at the point of consumption. Also, work force in this technique are locally available. The workforces are mostly the consumer themselves only. Running costs are almost negligible [3].

8. Disadvantages

This technology usually has a very limited number of limitations as follows. Rainwater harvesting technologies are mainly depending on rainfall which is uncertain in nature. This technology uses a "bottom-up" approach instead of the conventional "top-down" approach. Therefore, awareness about this technology is less. For a particular area, the type and size of rainwater harvesting are highly reliant upon the quantity and quality of its rainfall. Catchment area and type of catchment surface are also very vital. These structures cannot be used as the primary source of water supply because rainfall is unequally distributed in dif- ferent seasons in a year. Rainwater harvesting systems also depend on consumer demands and their affordable budget. The architecture of this system needs to design the structure considering every aspect from physical, environmental, to monetary point of view [26].

9. Cost

The building and maintenance cost of rainwater harvesting systems varies considering its catchment area, conveyance, and storing vessels. Place to place also, the cost varies. Like in a study, it is found that the cost of these structures in Asia varies from $0.17 to $0.37 per cubic meter of water storage. This cost is quite less than that of Africa. However, rainwater harvesting systems are less costly than the use of bore wells and tube wells if the initial investment does not include the cost of roofing materials. The lasting period of these structures is also more than 10 years. They need less maintenance and operation cost [27].

References

1. Yazar A, Ali A. Water harvesting in dry environments. In: Farooq M, Siddique K, editors. Innovations in Dryland Agriculture. Cham: Springer; 2016
2. Sharma RK. Rainwater Harvesting at National Institute of Technology (NIT) (doctoral dissertation). Rourkela; 2010
3. Kahinda JM, Taigbenu AE. Rainwater harvesting in South Africa: Challenges and opportunities. Physics and Chemistry of the Earth, Parts A/B/C. 2011;36(14-15):968-976
4. Bruins HJ, Evenari M, Nessler U. Rainwater-harvesting agriculture for food production in arid zones: The challenge of the African famine. Applied Geography. 1986;6(1):13-32
5. Kahinda JMM, Taigbenu AE, Boroto JR. Domestic rainwater harvesting to improve water supply in rural South Africa. Physics and Chemistry of the Earth, Parts A/B/C. 2007;32(15-18):1050-1057
6. Prinz D. Water Harvesting—Past and Future. In: Sustainability of Irrigated Agriculture. Dordrecht: Springer; 1996. pp. 137-168
7. Pretty J, Sutherland WJ, Ashby J, Auburn J, Baulcombe D, Bell M, et al. The top 100 questions of importance to the future of global agriculture. International Journal of Agricultural Sustainability. 2010;8(4):219-236
8. Che-Ani AI, Shaari N, Sairi A, Zain MFM, Tahir MM. Rainwater harvesting as an alternative water supply in the future. European Journal of Scientific Research. 2009;34(1):132-140

9. Kumar M, Kumar A, Vashisth A, Chhabra D, Sehgal D. A Project Report on Rainwater Harvesting at Lingaya's University; 2019

10. Shijila E. Development of a Filter System for Roof Water Harvesting (doctoral dissertation). Tavanur: College of Agricultural Engineering and Technology; 2014

11. Assessment of Best Practises and Experience in Water Harvesting, Rainwater Harvesting Handbook, African Development Bank, Tunis-Belvedere, Tunisia, 2013

12. Gould JE. Rainwater catchment systems for household water supply. In: Environmental Sanitation Reviews. ENSIC; 1991

13. Gould JE, McPherson HJ. Bacteriological quality of rainwater in roof and ground catchment systems in Botswana. Water International. 1987;12(3):135-138

14. Nissen-Petersen E. Rain Catchment and Water Supply in Rural Africa: A Manual. London: Hodder and Stoughton, Ltd.; 1982

15. Pacey A, Cullis A. Rainwater Harvesting: The Collection of Rainfall and Runoff in Rural Areas. London: WBC Print Ltd.; 1989

16. Schiller EJ, Latham BG. A comparison of commonly used hydrologic design methods for rainwater collectors. International Journal of Water Resources Development. 1987;3(3)

17. UNEP [United Nations Environment Programme]. Rain and Storm Water Harvesting in Rural Areas. Dublin: Tycooly International Publishing Ltd.; 1982

18. Wall BH, McCown RL. Designing Roof Catchment Water Supply Systems Using Water Budgeting Methods. Water Resources Development. 1989;5:11-18

19. Kinkade-Levario H. Design for Water: Rainwater Harvesting, Stormwater Catchment, and Alternate Water Reuse. 1st ed. New York: New Society Publishers; 2007

20. Malesu MM, Oduor AR, Odhiambo OJ, editors. Green water management handbook: Rainwater harvesting for agricultural production and ecological sustainability. SearNet Secretariat: World Agroforestry Centre; 2007

21. Mwangi JK, Shisanya CA, Gathenya JM, Namirembe S, Moriasi DN. A modeling approach to evaluate the impact of conservation practices on water and sediment yield in Sasumua Watershed, Kenya. Journal of Soil and Water Conservation. 2015 Mar 1;70(2):75-90

22. Carter DC, Miller S. Three years experience with an on-farm macro- catchment water harvesting system in Botswana. Agricultural Water Management. 1991 Apr 1;19(3):191-203

23. Snobar BA, Denis P, Suwwan MA. A plastic greenhouse for semi-arid agriculture (JAP greenhouse). Dirasat. 1988

24. Bozkurt S, Yazar A. Effects of different drip irrigation levels on yield and some agronomic characteristics of raised bed planted corn. African Journal of Agricultural Research. 2011 Oct 19;6(23):5291-5300

25. Imbira J. Runoff waterharvesting for crop production in Semi Arid Areas of Baringo. In: Thomas DB, Biamah EK, Kilewe AM, Lundgren L, Mochoge BO, editors. Soil and

Water Conservation in Kenya. Proceedings of the third National Workshop, Kabete, Kenya. 1989

26. Borthakur S. Traditional rain water harvesting techniques and its applicability. Indian Journal of Traditional Knowledge. 2009;8(4):525-530
27. Bhattacharya S. Traditional water harvesting structures and sustainable water management in India: A socio- hydrological review. International Letters of Natural Sciences. 2015;37:30-38

Section 3
Mapping of Smart Urban Development

Chapter-7

Mapping Smart Mobility Technologies at Istanbul New Airport Using the Customer Journey

1. Introduction

Today, 55% of the world's population lives in urban areas, and this proportion is expected to reach 68% by 2050 [1]. City governments face the challenge of produc- ing wealth, innovation, health, and sustainability [2], but with an increase in urban populations, these challenges become harder for city governments to manage. For this reason, instead of continuing to use old-fashioned methods, some cities have applied updated, innovative methods to manage such issues in smarter ways [3], much as we do in our daily lives. We have moved from using "dumb" technologies (e.g., a road atlas or telephone) to interacting with "smart" technologies (e.g., personalized journey planning apps on smart phones) that "exist to help us, serve us, to make our lives easier and more interesting" [4].

As internationalization has become a widely discussed topic in recent years [5], the problem of overcrowding of cities is considered a global issue, and smart city applications have spread rapidly worldwide [2]. These applications not only enable city governments to facilitate the routine functions of individuals, buildings, and traffic systems but also enable them to monitor, understand, analyze, and plan cities to improve their efficiency, equity, and quality of life in real time [6]. The concept "smart," therefore, means the ability to manage the complexity through Big Data, which comes from a variety of sources in a huge volume and in a rapid way.

As an example, in their daily lives, individuals check their social media accounts from mobile devices and share or like posts. They also read personalized news and check traffic jams. They then check their e-mail when they arrive at their offices. At lunchtime, they order food using online applications, and they watch personalized videos during free time. In other words, individuals produce a great deal of personal data and consume many more services based on that data. Thus, they also desire cities that support this lifestyle [7].

For instance, in Milton Keynes, a large town in the middle of England, the rapid expansion of the Internet together with young and technically inclined population gave rise to the concept of "smart cities." This included the application of tech- nology, especially Big Data principles, to improve residents' quality of life. Such applications ranged from waste management to public transportation [8]. Now, many more cities, such as London, Stockholm, Amsterdam,

Vienna, Luxembourg, Turku, Eindhoven, and Montpellier, have adopted a "smart" approach. The smart city theme also provides a city more stable and higher brand equity than green and creative themes [9]. In a project conducted at the Vienna University of Technology, six indicators of smart cities were identified:

1. "Smart economy," including competitive components, such as innovation, entrepreneurship, and productivity
2. "Smart people," which is the result of ethnic or social diversity and related to online education to raise social capital and qualification
3. "Smart governance," such as e-government, which can include all parties in transparent decision-making processes
4. "Smart mobility," which includes transportation planned using information and communications technologies (ICT)
5. "Smart environment," which encourages cities to use energy efficiently by employing innovative technologies, such as solar energy and other renewable sources
6. "Smart living," which refers to systems that improve quality of life, such as services and public safety tools [10]

These components are associated with different aspects of urban life: (1) industry, (2) education, (3) e-democracy, (4) logistics and infrastructure, (5) efficiency and sustainability, and (6) security and quality, respectively [11, 12].

This study is focused on smart mobility and examines the case of Istanbul New Airport, which officially opened on 29 October 2018. With a capacity of 200 million annual passengers and 3500 flights per day, Istanbul New Airport is the first smart airport of its size [13].

2. Bases of smart mobility

As an indicator of smart cities in terms of logistics and infrastructure, smart mobility requires urban planning that shifts the focus of transportation modes from individual to collective through the extensive use of ICT [14]. Smart mobility is concerned with local and global accessibility, ICT infrastructure, and innovative and sustainable transport systems [15]. Moreover, it should serve individuals' needs by reducing time spent traveling and helping travelers avoid unnecessary travel altogether [16]. To provide these outputs, smart mobility should [17] accomplish the following goals:

1. Use technology to generate and share data, information, and knowledge that influence decisions.
2. Use technology to enhance vehicles, infrastructure, and services.
3. Derive improvements for transport system operators, users, and shareholders.

In certain situations, IT-supported service experiences can enhance customer satisfaction significantly [18]. Therefore, customer journey maps can be analyzed to observe the effects of technologies on customers' experiences with the provided services.

3. Customer experiences enhanced with technology

Shorter life cycles make products and services more commoditized. Therefore, the differentiation has shifted from the offerings themselves to their providers that create experiences related with the acquisition, use, and maintenance of these offerings. People are naturally inclined to prefer pleasant, special experiences that have important lasting effects [19]. Thus, the opportunity for new revenue growth depends not only on driving sales of existing goods and services but on creating experiences for which customers are willing to pay [20]. Experiences can be conceptualized as "events that engage individuals in a personal way" [21] or as "enjoy- able, engaging, memorable encounters for those consuming these events" [22], so creating strong customer experiences is a leading objective for management [23] to create competitive advantage [24].

Given the current progress of technology, it is possible to use it to develop customer experiences. For instance, in the fashion industry, information technologies enhance customer experiences by creating interactive and exciting shopping experiences [25]. In the banking industry, the increase of the Internet services and automated teller machines (ATMs) in various locations provides more services and more comfortable customer experiences than before [26]. In addition to B2C context, technology is used also to create B2B customer experiences, such as sales- people's use of information technology [27].

In addition to enhancing experiences, technology advancement also provides multiple ways for customers to interact with product and service providers. These interactions are crucial to creating superior customer experiences [28]. To depict the events through which customers may interact with a service organization, aca- demics and practitioners use a strategic management tool called customer journey mapping (CJM) [29].

4. Customer journey mapping at Istanbul New Airport

Cities are considered smart when "investments in human and social capital and traditional (transport) and modern (ICT) communication infrastructure fuel sustainable economic growth and a high quality of life, with a wise management of natural resources, through participatory governance" [30]. As a dimension of smart city planning, smart mobility is considered in this study. Smart mobility can be summarized as planning and controlling transport systems through the extensive use of ICT. This kind of system has recently been applied in the building of Istanbul New Airport.

Airport information systems are divided into seven sections [31]:

- Flight planning and operation
- Passenger process
- Business administration
- Security
- Facility management
- Business center and airport management
- Contact and information

Recently, airports in the Middle East and in Istanbul have begun to compete with European airports to create a "global hub" for connecting (transfer) passengers [32]. With the increased capacity and technologies developed at Istanbul New Airport, an improved customer experience would provide a competitive advantage. Thus, CJM is proposed in this study to depict customer experiences based on the technologies adapted at Istanbul New Airport.

The format used by Rosenbaum et al. was considered when creating the CJM [29]. On the CJM's horizontal axis, the customer touch points take place according to a process timeline. The timeline is also divided into three periods: pre-service, service, and post-service. The pre-service period refers to customer experiences that occur before a service begins. The service period refers to customer experiences during the actual service. Finally, the post-service period refers to customer experiences occurring after the service [29]. In parallel with this process, a CJM can represent customer experiences prior to going to the airport, at the airport, and after leaving the airport, respectively.

4.1. Horizontal axis of the customer journey map

To develop a CJM for Istanbul New Airport, customer touch points are deter- mined first to build the CJM's horizontal axis. The general limitation of CJM indicated by Rosenbaum et al. was that a common underlying assumption for customer touch points was the consideration that each touch point was equally important to the planning process; however, not all customers experience all touch points [29]. Thus, 62 students with previous flight experiences were recruited from two undergraduate marketing classes at Marmara University using a convenience sampling method. As the Istanbul New Airport was not operating at full capacity at the time of the study, the selection criterion was limited to participants' previous flight experience at other airports in Istanbul (Istanbul Atatürk Airport or Istanbul Sabiha Gökçen Airport). To determine a unique CJM, only flights departing from the airports were considered. The main assumption for this approach was that passengers spend more time—or, in other words, they experience more—at the departure airport.

Table 1. Istanbul New Airport's key horizontal axis customer touch points.

	(% listed)
Pre-service period (before passengers arrive at the airport)	
Please list your pre-service (before going to the airport) touch points with the airport (such as "seeing an advertisement on a street billboard about the airport")	
1. Preparing luggage	1.61
2. Buying tickets (online or through agencies)	37.10
3. Transportation (public or private) to the airport	41.94
4. Checking in online	9.68
5. Visiting the airport's website or mobile application	9.68
6. Searching for information about facilities at the airport	1.61
7. Seeing advertising (on a street billboard, a website, or a social media service) about the airport	40.32
Service period (when passengers are at the airport)	
Please list your during-service (when you are at the airport) touch points with the airport (such as "using the parking lot of the airport")	
1. Using the parking lot	16.13
2. Security screening process	9.68
3. Checking in at the airport	19.35
4. Delivering luggage	8.06

5. Asking for help from airport personnel (such as asking for general information or for a wheelchair)	1.61
6. Eating or drinking in the cafés at the airport (including smoking rooms)	56.45
7. Using an ATM to withdraw money	1.61
8. Exchanging money	1.61
9. Using lounge services	9.68
10. Using the toilets	9.68
11. Shopping at the airport	11.29
12. Paying for stamp fees	3.23
13. Passport control	0
14. Tracking flight gates on the screens	1.61
15. Going to flight gates	4.84
16. Visiting duty-free stores	11.29
17. Waiting in the waiting areas (such as reading a book or listening to music)	16.13
18. Connecting to Wi-Fi	8.06
19. Taking photos in the airport	1.61
20. Boarding	3.23
Post-service period (after passengers leave the airport)	
Please list your after service (when you leave the airport) touch points with the airport (such as "talking to friends or family about the airport")	
1. Baggage collection (at the destination airport)	9.68
2. Calling (informing) family about arrival	12.90
3. Talking to friends or family about experiences at the airport	48.39
4. Communicating with the airport post-services (completing a survey about experiences at the airport)	3.23
5. Writing about the airport on the Internet	1.61
6. Sharing photos taken at the airport on social media	1.61

As discussed previously, the main assumption of CJM is that each passenger's touch points are considered to be of equal importance. To address this limita- tion, the students were asked to indicate the first touch points that came to mind regarding the departure airports in Istanbul. As they all experienced different touch points, all the steps gathered from the students are listed in **Table 1** to create a complete list of customer journey touch points. The percentages of listed touch points are indicated next to them. At the end of this step, a total of 33 touch points was identified by the respondents. Of the steps, 7 occurred during the pre-service period (before passengers arrive to the airport), 20 occurred during the service period (when passengers are at the airport), and 6 occurred during the post-service period (after passengers leave the airport). After the collection of the touch points for Istanbul Airport as a departure airport, two marketing professors from Marmara University checked them for a potential missing point.

4.2. Vertical axis for corresponding technological applications

The vertical axis for CJM reflects the managerial practices that enable passengers to experience each touch point in a satisfactory way [29]. As the purpose of this study is to reveal customer experiences with technology applications, only the corresponding technological application is considered regarding passenger touch points at Istanbul New Airport.

Turkish Airlines' move from Istanbul Atatürk Airport to the Istanbul New Airport coincides with the time of this study (5–7 April 2019), so it was not possible to meet with authorities at Istanbul New Airport. However, secondary data were used to list the corresponding technological applications. A meeting was held with Ms. Hülya Zerener

Gürbaşak, the corporate account manager of Technopc, which provides more than 4000 hardware products to Istanbul New Airport. These products include desktop, mini, integrated, industrial, and kiosk computers used at check-in, security, card access, and passport control points. In addition, Mr. Mertcan Tanaydý, the communication chief of ÝGA Istanbul Airport, provided several documents and a video [33] in which the tech- nologies of the airport are explained in detail from a customer's point of view.

In terms of technological infrastructure at Istanbul New Airport, three data centers support artificial intelligence (AI) and smart systems. There are 647 servers, 3267 flight displays, 4549 computers, and more than 3000 card access points. To integrate these systems, 1740 km fiber and 4500 km copper cables were used. Data from all systems can reach up to 209 million GB. A total of 708 employees works in technology support roles [33].

The main focus of the technology was a mobile application developed to assist passengers. The airport mobile application can be used before, during, and after service. It guides passengers from home to the airport, assisting with time management by considering the traffic on the road (corresponding to the pre- service touch point). Its "Where is my car?" service also helps passengers find their cars in the parking lot among 19,000 cars in the closed area and 40,000 cars in the open area (corresponding to the post-service touch point). All corresponding technological applications for the service period are listed in **Table 2**.

Table 2. Corresponding technological applications during the service period at Istanbul New Airport.

Passenger touch points during the service period	**Corresponding technological applications**
1. Using the parking lot	A total of 4500 cameras tracks the cars in the parking lot; they take photos of the cars and upload them on the airport mobile application. Visitors may find their cars by writing their license plate on the application
2. To be controlled by security	Electronic screens at the airport entrance show wait times at the security points (based on sensors and cameras that produce heat maps showing the density of people). Accordingly, the number of security points increases or decreases. In addition, the security cameras with AI use facial recognition systems and warn security if they detect unfavorable movements in the airport

3. Checking in at the airport	Self-check-in points allow passengers to check in for their flights and leave
4. Delivering luggage	
5. Asking airport personnel for help (asking for general information or for a wheelchair)	An interactive passenger assistant located at the airport enables passengers to connect to customer service and make a video call with a responsible party who can see the passenger's flight details; the customer opens the e-ticket's QR code and puts the mobile phone to the device. The cameras and sensors on the top scan the e-ticket's QR code, enabling the responsible party to see both the passenger and the ticket. In this way, the responsible party can generate a personal map for the passenger from the current location to the flight gate
6. Eating and drinking in the cafés at the airport (including lounge services and smoking rooms)	The airport location guide provided by the airport mobile application lists all facilities and stores at the airport
7. Using an ATM to withdraw money	
8. Exchanging money	
9. Using lounge services	
10. Using the toilets	
11. Shopping at the airport	
12. Paying for stamp fees	
13. Passport control	The iGate-fast passport control system includes two steps. The first door is opened when the passport is scanned by a device, and the second door is opened when a passenger's face and hand are scanned by a device and matched to the passport
14. Tracking flight gates on screens	Flight information screens and the airport mobile application provide flight information
15. Going to flight gates	Personalized direction is provided by the airport mobile application to the gate for the selected destination
16. Visiting duty-free stores	Pre-ordering is available in duty-free stores, where salespeople can reserve passengers' pre-ordered items
17. Waiting in the waiting areas (reading a book and listening to music)	USB inputs between the seats enable passengers to charge their mobile devices
18. Connecting to Wi-Fi	Up to 1 h of free Wi-Fi is available at the airport
19. Taking photos in the airport	
20. Boarding	Boarding time can be seen in the airport mobile application

In addition to existing technologies, new ones are planned for the airport. There will be a store offering inventions and technological products where visitors may have such experiences as flying a drone. Moreover, three types of robots will work in the airport. Service robots will assist passengers and will reply to their questions about such topics as flight gates and ticket offices. The other robots will be responsible for cleaning and safety. They will be called "ÝGAbots." Another innovation will be autonomous vehicles, which will be called "ÝGAbus." They will carry passengers to various places at the airport [33].

5. Conclusion

City governments today must manage increasing urban populations in smart ways. In other words, large amounts of data produced by the population should be used to create sustainable places by decreasing the chaos of overcrowding, such as traffic, pollution, and waste. One smart city dimension is smart mobility, which refers to enhancing individuals' mobility using innovation and technology. Technology is also essential to creating customer experiences. CJM can be used to analyze touch points and their corresponding created experiences.

In this study, an attempt was made to use CJM as a tool to analyze Istanbul New Airport, one of the largest smart airports in the world, which promises memorable customer experiences. In this way, Istanbul New Airport aims to compete with other airports defined as hubs in Europe. Therefore, a customer journey was cre- ated based on a survey of 62 undergraduate students who were asked which touch points came to mind first. The corresponding technologies were indicated based on secondary data.

According to the results, the most indicated touch point from the participants was "talking to friends or family about experiences at the airport" (48.39%).

However, the question should be considered to evaluate this high rate. To make the questions clear and to get more touch points, an example was provided for preservice, service, and post-service periods. Therefore, these three questions were recalls rather than the points coming to the respondents' minds first. Nevertheless, the findings provide evidence that touch points do not have equal importance for all passengers. Different passengers experience different touch points during their journeys, so they perceive different experiences with the technological applications.

As can be seen from the touch point-technology match, Istanbul New Airport provides experiences through technological applications for most passenger touch points. Of these technologies, the main focus is on the mobile application created to affect various points of a customer journey, including time management and considering the traffic on the road to the airport, car location in the parking lot, guides listing all the facilities and stores in the airport, flight information for the selected destination, and personalized directions to the flight gate. In addition, various technologies are used to enhance passenger experiences, such as self-check-in points, interactive passenger assistants, AI-integrated security cameras, fast passport control system, and free Wi-Fi. According to the CJM, customer experience creators should also consider passengers' use of social media and create a strategy accordingly. Passengers like to take photos at the airport and share them on their social media accounts, so there may be several photo opportunities when they experience a new technology. In this way, passengers may share and transfer their experiences with their surroundings, leading to positive word-of-mouth communication about the airport.

Regarding the limitations of this study, as the airport was not running at full capacity, the CJM was created based on participants' experiences at other airports in Istanbul. Therefore, a new CJM should be created after the airport begins to run at full capacity by considering the airport from both departure and arrival perspec- tives to extend customers' experiences. In addition to technological applications, other managerial practices can be also included in the CJM's vertical axis.

References

1. United Nations. Revision of World Urbanization Prospects [Internet]. 2018. Available from: https://www.un.org/ development/desa/publications/2018- revision-of-world-urbanization- prospects.html
2. Meijer A, Bolívar MPR. Governing the smart city: A review of the literature on smart urban governance. International Review of Administrative Sciences. 2016;82:392-408. DOI: 10.1177/0020852314564308
3. Nam T, Pardo TA. Smart city as urban innovation: Focusing on management, policy, and context. In: Proceedings of the 5th International Conference on Theory and Practice of Electronic Governance (ICEGOV '11); 26-29 September 2011; Estonia. New York: ACM; 2011. pp. 185-194
4. Brenner S. Law in an Era of Smart Technology. New York: Oxford University Press; 2007. p. 194
5. Philippi C, Bobek V, Macek A, Justinek G, Horvat T. Internationalization of an Austrian SME with a sales agent to Mexico and the USA in the automotive sector from cost perspective. International Journal of Globalisation and Small Business
6. Batty M, Axhausen KW, Giannotti F, Pozdnoukhov A, Bazzani A, Wachowicz M, et al. Smart cities of the future. European Physical Journal. 2012;214:481-518. DOI: 10.1140/epjst/ e2012-01703-3
7. Dirsehan T. Akýllý Ţehirlerin Yerel Yönetimleri (Local Governments of Smart Cities). Harvard Business Review. Türkiye Blog [Internet]. 2019. Available from: https:// hbrturkiye.com/ blog/akilli-sehirlerin-yerel-yonetimleri
8. Marr B. Big Data in Practice. UK: Wiley; 2016. 308 p
9. Chan C-S. Which city theme has the strongest local brand equity for Hong Kong: Green, creative or smart city? Place Branding and Public Diplomacy. 2019;15:12-27. DOI: 10.1057/ s41254-018-0106-x
10. Giffinger R, Fertner C, Kramar H, Kalasek R, Pichler-Milanovic N, Meijers E. Smart Cities—Ranking of European Medium-Sized Cities (Report). Vienna University of Technology; 2007. Available from: http://www.smart-cities.eu/ download/smart_cities_final_report.pdf
11. Lombardi P, Giordano S, Farouh H, Yousef W. Modeling the smart city performance. Innovation: The European Journal of Social Science Research. 2012;25:137-149. DOI: 10.1080/13511610.2012.660325

12. Albino V, Berardi U, Dangelico RM. Smart cities: Definitions, dimensions, performance, and initiatives. Journal of Urban Technology. 2015;22:3-21. DOI: 10.1080/10630732.2014.942092

13. Şahin A. Ýstanbul Yeni Havalimaný ile dünya daha akýllý (The world is smarter with Istanbul New Airport). Fortune. Turkey [Internet]. 2017. Available from: http://www.fortuneturkey.com/istanbul-yeni- havalimani-ile-dunya-daha-akilli-49928 [Accessed: 24 February 2019]

14. Letaifa SB. How to strategize smart cities: Revealing the SMART model. Journal of Business Research. 2015;68:1414-1419. DOI: 10.1016/j. jbusres.2015.01.024

15. Chauhan S, Agarwal N, Kar AK. Addressing big data challenges in smart cities: A systematic literature review. Info. 2016;18:73-90. DOI: 10.1108/ info-03-2016-0012

16. Stone M, Knapper J, Evans G, Aravopoulou E. Information management in the smart city. The Bottom Line. 2018;31:234-249. DOI: 10.1108/ BL-07-2018-0033

17. Lyons G. Getting smart about urban mobility—Aligning the paradigms of smart and sustainable. Transportation Research Part A. 2018;115:4-14. DOI: 10.1016/j. tra.2016.12.001

18. Dawes J, Rowley J. Enhancing the customer experience: Contributions from information technology. Management Decision. 1998;36:350-357. DOI: 10.1108/00251749810220568

19. Carbone LP, Haeckel SH. Engineering customer experiences. Marketing Management. 1994;3:9-19

20. Gilmore JH, Pine BJ II. Customer experience places: The new offering frontier. Strategy & Leadership. 2002;30:4-11. DOI: 10.1108/10878570210435306

21. Pine IIBJ, Gilmore JH. The Experience Economy: Work is Theatre & Every Business a Stage. Boston, MA: Harvard Business School Press. 272 p

22. Oh H, Fiore AM, Jeoung M. Measuring experience economy concepts: Tourism applications. Journal of Travel Research. 2007;46:119-132. DOI: 10.1177/0047287507304039

23. Lemon KN, Verhoef PC. Understanding customer experience throughout the customer journey. Journal of Marketing. 2016;80:69-96. DOI: 10.1509/jm.15.0420

24. Edelman DC, Singer M. Competing on customer journeys. Harvard Business Review. November 2015:88-100. Available from: https://hbr.org/2015/11/ competing-on-customer-journeys

25. Blázquez M. Fashion shopping in multichannel retail: The role of technology in enhancing the customer experience. International Journal of Electronic Commerce. 2014;18:97-116. DOI: 10.2753/JEC1086-4415180404

26. Aliyu AA, Bin R, Tasmin HJ. The impact of information and communication technology on banks' performance and customer service delivery in the banking industry. International Journal of Latest Trends in Finance and Economic Sciences. 2012;2:80-90

27. Agnihotri R, Dingus R, Hu MY, Krush MT. Social media: Influencing customer satisfaction in B2B sales. Industrial Marketing Management. 2016;53:172-180. DOI: 10.1016/j. indmarman.2015.09.003

28. Barwitz N, Maas P. Understanding the omnichannel customer journey: Determinants of interaction choice. Journal of Interactive Marketing. 2018;43:116-133. DOI: 10.1016/j. intmar.2018.02.001

29. Rosenbaum MS, Otalora ML, Ramírez GC. How to create a realistic customer journey map. Business Horizons. 2017;60:143-150. DOI: 10.1016/j.bushor.2016.09.010

30. Caragliu A, Del Bo C, Nijkamp P. Smart cities in Europe. Journal of Urban Technology. 2011;18:65-82. DOI: 10.1080/10630732.2011.601117

31. Istanbul New Airport Guide [Internet]. Available from: https:// www.igairport.com/ tr/igadocuments/ havalimanirehberitr.pdf

32. Oxera. The continuing development of airport competition in Europe [Internet]. 2017. Available from: https:// www.oxera.com/publications/the- continuing-development-of-airport- competition-in-europe/

33. CNN Türk. Teknoloji Her Yerde (Technology is Everywhere). 16.02.2019. Uploaded by ÝGA Istanbul Airport. [Internet]. Available from: https://www. youtube.com/ watch?v=aFI5lcVgDz0

Chapter-8

Environmental Noise Mapping as a Smart Urban Tool Development

1. Introduction

The European Environmental Noise Directive 2002/49 [1] implemented in all EU Member States, almost 18 years ago, provided several smart tools to access and manage environmental noise and enhance cities' development. As the directive stipulates, "it is part of community policy to achieve a high level of health and envi- ronmental protection, and one of the objectives to be pursued is protection against noise. In the Green Paper on Future Noise Policy, the Commission addressed noise in the environment as one of the main environmental problems in Europe" [2].

This chapter offers, therefore, an analysis of the tools that have been created in the framework of this directive and aims to show how these specialized tools con- tribute to an intelligent development of European cities and wider urban territories. The analysis is based on a series of practical cases studies carried out in Greece (and in Europe) and will show how these smart tools had to adapt to the twenty-first- century environmental issues.

2. The environmental noise directive as a "toolbox" for smarter cities

The environmental noise directive concerns, since its enforcement in both European and national framework, not only the major agglomerations in EU Member States but also all main transportation infrastructure. Environmental noise is defined by the traffic noise from road, railway, and airport infrastructures combined with industrial sources. Each Member State has incorporated this directive, into its national legislative framework and therefore has the obligation to implement it in their relevant urban agglomerations and territories. This was an important step forward for the environmental noise both on national and European scales, because it created the appropriate framework for policy-makers, politicians, transportation engineers, urban planners, architects, and also every citizen to share information and interact on the definition of all appropriate regulation and mitigation measures. In other words, it allowed the Europeans to address in the same language their concern on the environmental noise issue. A strategic noise map (SNM) is therefore primarily requited allowing to visualize the "decibel" impact of the main sources of environmental noise either at the scale of an agglomeration or at the scale of an transportation or industrial infrastructure. As the next step, a comprehensive noise action plan (NAP) is therefore drawn with the involvement of transportation, planning, and acoustic engineers to access and specify the most appropriate means to achieve the needed noise rehabilitation by mitigation measures. The general framework and the basic homogenous methodology have been applied in the majority of the Member States, and now, 18 years later, we can see the important advantages emerging.

2.1. Smart tools for environmental noise measurement issue

As analyzed above, in this same legislative framework, transportation opera- tors needs to measure and simulate the noise impact on the environment. The noise emitted by a vehicle or an airplane is a dynamic source that evolves in time and space. The environmental noise emitted by the traffic of the road traffic flow (including motorcycles and trucks) over a day is therefore an inexhaustible source of information that continues to evolve (big date issue) and which poses questions to traffic engineers for its measurement, its prediction, and furthermore its management. The European Directive introduces several smart tools in order to solve these problems.

For example, Attiki Odos, the road operator for Athens Ring Road has been awarded in 2003 (Decibel d'Or, Ministère de l'Environnement en France) for its monitoring system with eight permanent monitoring stations that measure in real time road noise traffic (see **Figure 1**). In Attiki Odos, since its opening in 2004 for the Olympic Games, more than 250,000 vehicles are passing every day [3].

In order to achieve this goal, Members States are applying the directive by using the common indices L_{den} and L_{night}. The European Directive therefore is applied to environmental noise to which humans are exposed by introducing the above noise indicators that shall be determined by homogenous assessment methods. The definition of the L_{den} level (day-evening-night) is defined by the following formula:

$$L_{den} = 10 \lg \frac{1}{24}\left(12 \times 10^{\frac{Lday}{10}} + 4 \times 10^{\frac{Levening+5}{10}} + 8 \times 10^{\frac{Lnight+10}{10}}\right) \quad (1)$$

where L_{den} is expressed in A-weighted decibel or dB(A); L_{day} is the A-weighted long-term average sound level as defined in ISO 1996-2: 1987, determined over all the day periods of a year; $L_{evening}$ is the A-weighted long-term average sound level as defined in ISO 1996-2: 1987, determined over all the evening periods of a year; L_{night}

Figure 1. Permanent noise and air pollution monitoring systems deployed on Attiki Odos Ring Road, Athens, in use from 2001 to 2002: view of typical measurement stations and the CUBE measurement system (Dynacoustics & 01dB-ACOEM).

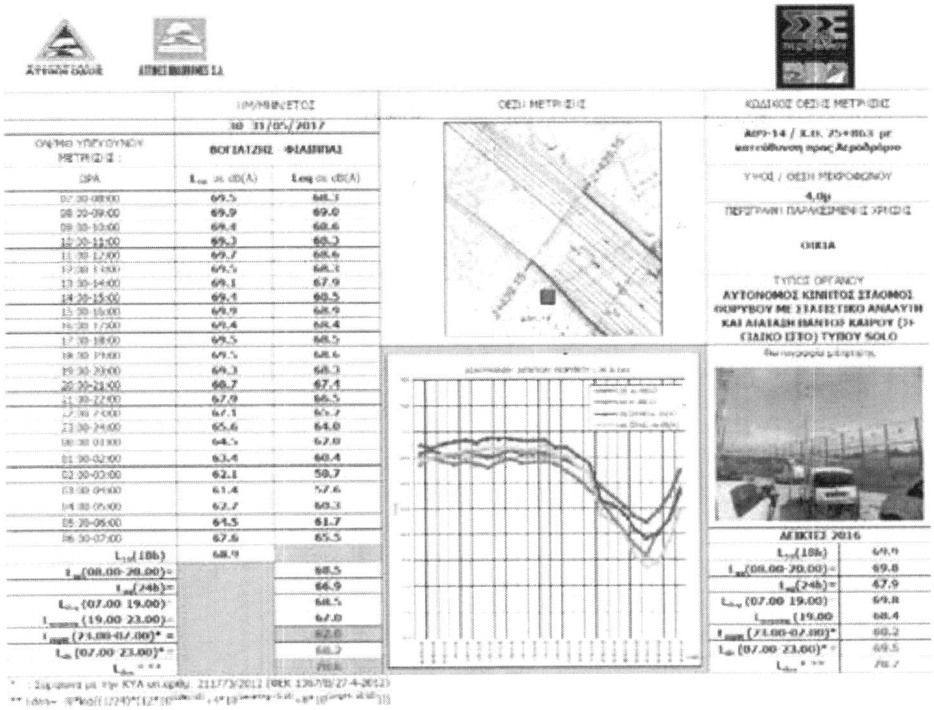

Figure 2. A typical 24-h measurement in Athens Ring Road (2017) (accessed in three periods: day, evening, and night).

is the A-weighted long-term average sound level as defined in ISO 1996-2: 1987, determined over all the night periods of a year.

As a result, 24-h measurements on Athens peripheral can be presented as follows:

As presented, in **Figure 2**, these tools allows to sum up 1-year measurement in few values that takes into count the density of the road traffic depending on the hours of the day, the week, and the month. The formula shows that the road traffic is even more annoying at night and in the evening than at the hours of the day. Thatis why the formula introduces a weight system that gives more emphasis to noise sources that appears during the evening and during the night: a penalty of 5 dB(A) for the evening and for 10 dB(A) for the night period (see Eq. (1)).

It is one of the main smart tool since it can give at a specific point one value for a big data problem: if we take a measurement each second, it summarizes 1 year of measurement that is to say 60 s × 60 min × 24 h × 365 days, for example, 31,536,000 s of potential measurement period per year. Practically, these measure- ments based on 24-h periods can be repeated and used to calibrate acoustic models in order to simulate with the best accuracy the environmental noise propagation (from a road section scale to the whole agglomeration scale). For example, a part of the strategic noise map of Athens simulated on CadnaA software and calibrated with 24-h measurements is presented in **Figure 3**.

Scientific papers published by several teams have shown comparison from real in situ L_{den} 24-h measurements and the one simulated has a correlation index more than R > 0.91 [5] (see **Figure 4**).

The measurement indices, therefore, can be simulated with high precision, and in its average, it resists the qualitative and quantitative variations of the yearly average traffic. Another word, with the above strategic noise map correlation methodology, measurements and simulations of the noise indices are energetically correct and express the quantity of noise than a monitoring station may record during the whole year at a specific point (something that is practically and financially impossible to do).

Figure 3. Part of the strategic noise map of Athens (2016) [4].

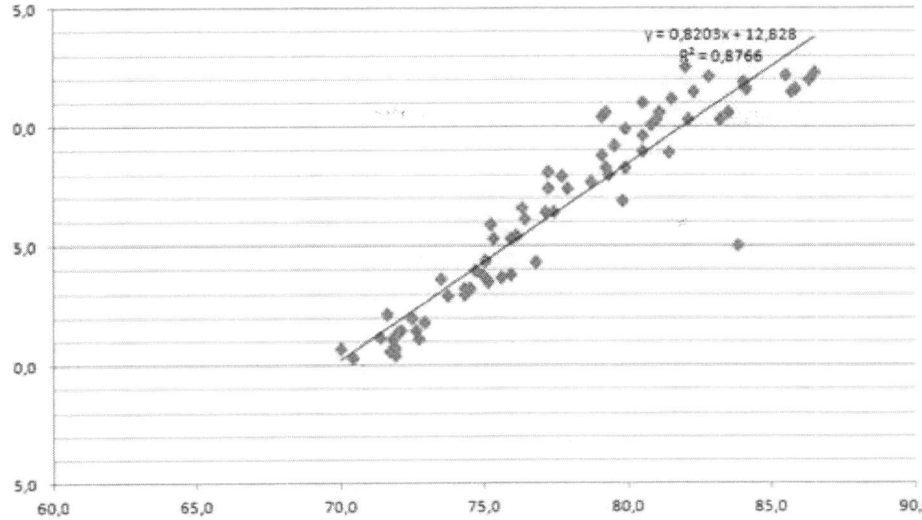

Figure 4. L_{den} **index (Athens SNM). Correlation measured vs. calculated levels** [5].

As a result, the use of both L_{den} and L_{night} indices appears to be an adequate and quite adapted even intelligent tool that allows transportation operators and urban municipalities to assess the environmental noise impact of the development of cities on a "long term" (yearly operation). In Europe, the majority of municipalities with more than 100,000 inhabitants and the major transport operators (cars, trains,

and planes) produced strategic noise maps, updated every 5 years, according to the directive; however their implementation has in several cases been delayed. Besides the fact that in some European countries the administration did not keep up with this pace, we need to underline that in some cases the early implementation of permanent surveillance systems was introduced (e.g., the Athens Ring Road and the Athens International Airport) [5, 6].

2.2. Smart tools for environmental noise exposure

As per the previous analysis, the relevant European acoustic criteria, which can be measured and predicted, are smart tools for expressing simply the amount of acoustic energy received at a point exposed to different sources of environmental noise. It is thus possible to edit strategic noise maps and link them to the relevant geographical information systems. Thus, these maps become strategic because they can easily express the amount of people exposed to different sound levels. These noise classes has been standardized (in dB(A) and a corresponding color in the map, as per **Figure 3** where the noise classes for L_{den} and the relative color code were used as standardized by European Directive 2002/49).

GIS systems can easily cross statistical inhabitants' localization with SNM and bring to the light the number of people exposed to several noise levels. Based on the European Directive-introduced noise indices, each Member State has the right to adapt in their national legal framework specific limits to define the level of noise pollution (see **Table 1**).

One might criticize the fact that not all European countries have the same limit values [7]. Indeed, as shown in **Table 1**, GR requirements are less demanding (therefore easier to reach) than those recommended by the European Union as a min population exposure level. This is rather a delicate subject that deserves some explanation since European standards are often proposed by Northern Europe Member States for which the economic and social development is often considered more advanced than for Southern Europe Member States. It is the intelligence of the directive and its criteria that gives the possibility of each Member State to adjust its limit levels according to its own geography, climate, lifestyle, social structure, and economy. The GR limit levels may seem extremely easier to achieve than the German ones. It must be understood, however, that Greece, Spain, or Southern Italy are characterized by Mediterranean climate conditions and lifestyle that are quite different from those of, for example, Berlin, Copenhagen, or Stockholm. The periods of the typical day expressing the levels of noise correspond rather to lifestyles of Northern and Western Europe than to Southern Europe. Although it is less and less the case in national capitals and large urban agglomerations (e.g., Madrid, Athens, Nicosia, Rome, or Naples), the peaks of activities are, for example, often 2–3 h after Paris relevant ones. The clipping 07:00–19:00 for L_{day}, 19:00–23:00 for $L_{evening}$, and 23:00–07:00 for L_{night} are not quite adaptable for Greece or Cyprus, for example. The evening in a GR city is maybe the noisiest period of the day, with GR people work- ing—in the private sector especially in commerce—until 21:00, and going for dinner toward 22:00 or even 23:00; therefore the relevant noise measurements will weigh more in the general formula of the L_{den}, as per the penalties introduced in Eq. (1).

Table 1. National regulations concerning maximum values of noise pollution indicators (i.e., for L_{den} and L_{night}) in Greece compared to European recommendations for population exposure [7].

Environmental noise levels	L_{den}	L_{night}
Greece (GR)	<70 dB(A)	<60 dB(A)
Europe (min exposure levels for SNM)	<55 dB(A)	<50 dB(A)

Following the European common methodology as per the latest update of the Annex II introduced recently by the European Directive 2015/996 [8], many agglomeration and transport operators present and share their results on the European portal of the European Environmental agency, the noise Observation and Information service for Europe [7]. Main results of most of the main cities and infrastructure of the majority of European members States member states are avail- able. With a simple "click," it is easy to get the following information: total of people exposed to noise from road traffic (but also railway and airport traffic are avail- able), during the day and during the night. It shows how much people are exposed to high level of noise and present graphics that describes the statistical partition of this exposure.

For example, in **Figure 5**, main results are presented for Amsterdam (Haarlem) and Berlin. The web site, trough popup windows, explained clearly how much people are disturbed by noise traffic in the two cities: almost 600,000 inhabit- ants for both cases with approximately 50,000 more for Berlin. These data, from the noise exposure point of view, are comparable because they describe the same family of criteria (L_{den}, L_{day}, L_{night}), because methodology to measure, calculate, or simulate these values is also standardized. Surely, such comparison is quite help- ful for law makers, in each country, to organize their policy for noise mitigation.

Noise issues, even if they concern almost the same amount of people in Berlin or in Amsterdam, cannot be dealt in the same way when one knows the specificity of each country concerning town planning, building density, urban sprawl, etc.

Even so the portal is missing information for some Member States, it is already a huge step forward in order to understand the noise issue at the national and European scale. Data are comparable (they use the same criteria), and the map representation of this data allows also transversal analysis regarding noise exposure, data traffic, and territorial properties.

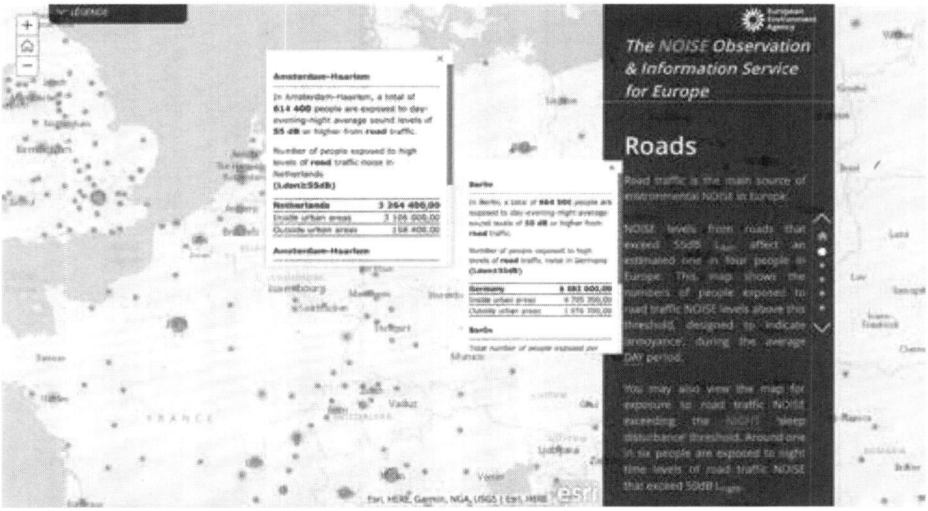

Figure 5. Visualization of the number of people exposed to high level of road traffic (more than L_{den} > 55 dB(A)) in Amsterdam (Haarlem) and Berlin agglomerations (source: European Environment Agency portal) (noise observation and information service for Europe web site—http://noise.eea.europa.eu).

In the same idea, airports have been mapped, and their strategic noise maps and have shown their influence on the city they border, as in Alikarnassos municipality close to Heraklion International Airport Nikos Kazantzakis in Crete in Greece [4]. The use of L_{den} and advanced prediction models calibrated with in situ 24-hour measurements allows to predict, with high precision, the environmental noise levels in any agglomeration. In this example, the main idea of an appropriate and effective noise action plan was to relocate the international airport from this area to a less built environment almost 20 km far away from the city center (project in execution stage).

In the example above, the comparison of the two SNM presents the impact of the environmental noise generated by the air traffic and especially its influence on the whole neighborhood studied here. This tool intelligently reinforces the scenario for moving the airport to a less developed area [9, 10]. Therefore by providing a common framework, the EU Member States have introduced intelligent tools that allow the simple translation and assessment of a large number of sources of environmental noise.

These tools are accessible to all the graphic representations, and the results are shareable with all the main decision-makers in a given agglomeration who ensure their participation in the decisions aiming to address the sustainable development of the acoustic environment of the cities.

2.3. Smart tools for city development

According to the European Directive, after the execution of SNM, appropriate noise action plans have been drawn in agreement with the existing and foreseen local policies. Those

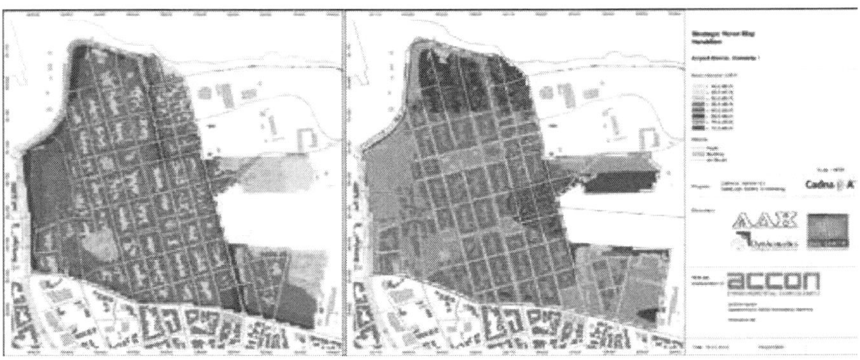

Figure 6. NAP for Alikarnassos district in Heraklion area adjacent to the International Airport Nikos Kazantzakis: on the left: L_{den} road and on the right: L_{den} road and air traffic [9, 10].

action plans as also the relevant SNMs are linked with geographical information system, so the smartness of the criteria is very much linked with its capacity to correlate the acoustic data with any data within the GIS database of the strategic noise maps and noise action plans, even if they are represented in two dimensions, with a common height level of 4 m. They are actually calculated in full three-dimensional geographical system in order to simulate properly the sound propagation in a complex city 3D environment. Tools have been built in order to integrate the exact topography of the relevant study area and of course the influ-

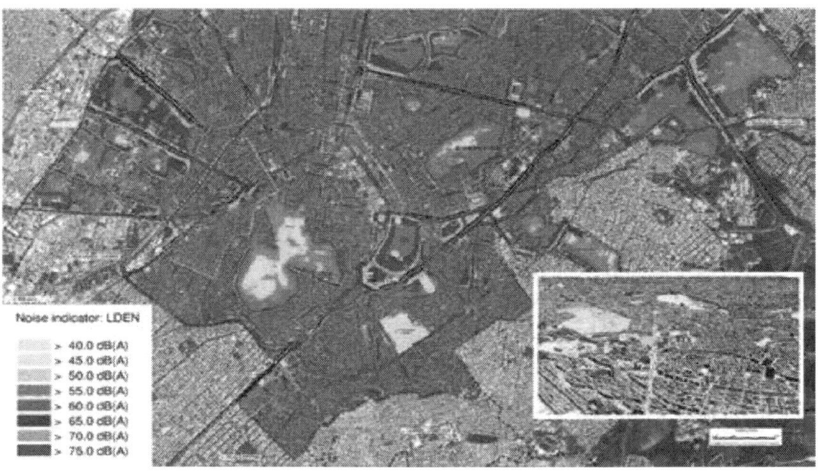

Figure 7. L_{den} strategic noise maps for South Athens agglomeration (2017) and partial view of the GIS 3D model.

ence of the built space on noise propagation (the building in its three dimensions including reflection characteristics). Therefore it is very efficient for all actors (politicians, mayors, town planners, engineers, acousticians) to represent the data on a map on a satellite view of the city, where a distinct color palette represents the accessed noise class (for every 5 dB) as per **Figures 4** and **6** and as also per the South Athens SNM relevant indexes hereafter (**Figure 7**).

By building the acoustic model on a complete GIS environment, it is also possible to calculate the exact number of people exposed to relevant levels of noise. It is also possible to zoom in the model and see if a specific building depending on its orientation is exposed or not to high level of noise. GIS contains a set of full data regarding the number and the geolocalization of the points of interest and sensitive receptors, for example, hospitals, education buildings, religion buildings, parks, and quite zones. So it is quite easy, after the superimposition the noise level contours on the GIS platform, to execute a quantitative analysis on the exposure factor. Previous studies have shown that, on Athens Ring Road (see **Figure 2**), noise exposure

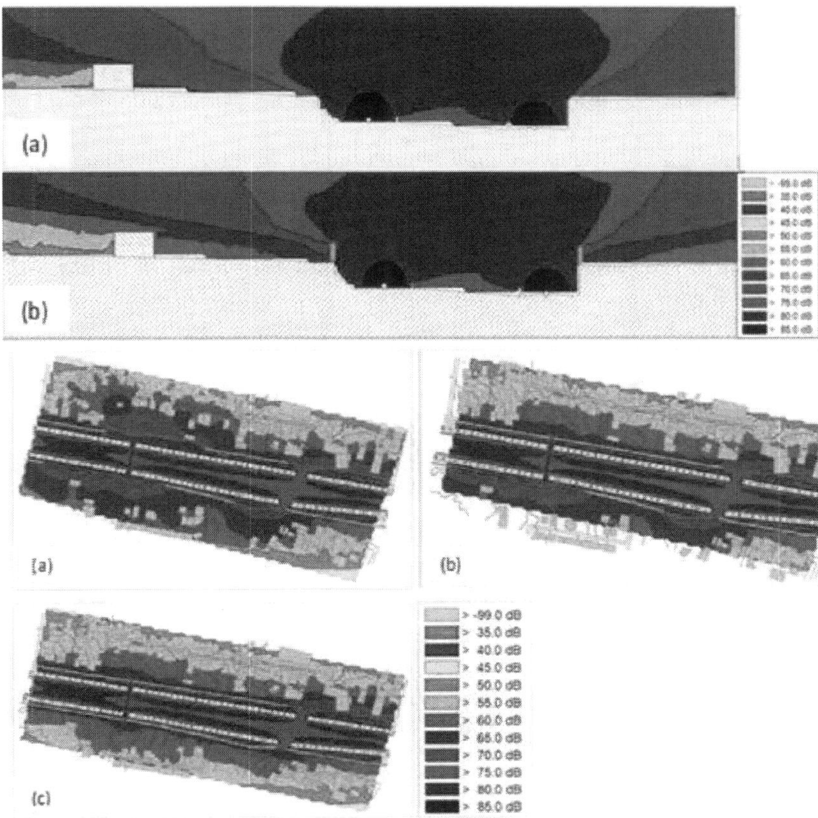

Figure 8. Athens Ring Road NAP (2017) [11]. Vertical grid—noise index L_{den}: (a) condition with no mitigation measures, (b) condition with the mounting of noise barrier. Horizontal grid—noise index L_{den}: (a) first scenario with no mitigation measures, (b) second scenario with barriers in two receptors, and (c) third scenario with barriers in three receptors (fully covered area).

needs to be monitored for more than 170 points of interest as hospital, clinics, maternity, childcare, education buildings, cultural uses, and worship places [4]. Thus, for a municipality, it is possible to prioritize and focus accordingly on appropriate public and private policies and launch mitigation programs aiming to improve the acoustic environment enveloping these sensitive receptors.

Similar conclusions were drawn specifically for the airport noise exposure. For example, in Heraklion, within the relevant NAP, specific studies were completed in order to access the cost for the acoustic insulation and the rehabilitation of both public and private buildings in the case of a "no-moving" scenario for the inter- national airport [4]. In this case the relevant costs of implementing an effective acoustic insulation regulation in buildings and maintaining the airport activity were calculated, in order to improve the acoustic environment of the district of Alikarnassos.

In this perspective, noise action plans have been accessed in order to minimize the population noise exposure. More than 3000 m^2 of noise barriers have been com- pleted on the Athens Ring Road during the last years [11]. Their implementation was based on the provisions of the European Directive 2002/49, and the GR legal framework introduced max thresholds for the relevant indices L_{den}, L_n, and L_{de}, for example, 70, 60, and 67 dB(A), respectively. When the statutory limits of the noise indices were exceeded, the implementation of immediate mitigation measures was suggested (e.g., noise barriers). The implementation of esthetic noise barriers with effective acoustic heights up to 4.5 m was proven successful and very well welcomed by the habitants. In order to resolve the issue of the environmental noise exposure on the population exceeding the existing criterion and limit, a full analysis of the implementation of adequate noise barriers was executed for all within the Noise Action Plan 2017 for the Athens Ring Road [11] (**Figure 8**).

The early surveillance of Athens' road traffic made it possible even during the early years of the operation to implement a comprehensive program for noise management and monitoring. As the construction of the motorway is at a level of −14 m below the ground level, very often, Attiki Odos was partially covered to reduce noise emissions and minimize the local residents' noise exposure. Most of the time, it was a good opportunity to introduce sports and social facilities (soccer fields, tennis courts, playgrounds, parks, etc.) and rehabilitate effectively the urban environment (**Figure 9**).

The development of an urban agglomeration is therefore strongly conditioned by the results of the noise monitoring and the implementation of the relevant smart tools ensuring effective keys to the decision-makers introducing appropriate measures. In another scale, the GR medium-sized city of Volos (approximately 120,000 inhabitants), on the east coast of Greece, has set up one of the main elements of its action noise plan. Indeed, the entrance of the city was a source of important road traffic noise because of continuous congestion effects in selected intersections. The NAP proposed the installation of five one-level roundabouts in order to streamline traffic, reduce traffic speeds, and thus reduce the noise emitted by vehicles. In the late 2018, four roundabouts are already constructed, and local studies and monitor- ing programs have already shown the positive impact of these mitigation measures concerning the environmental noise [9] (**Figure 10**).

Another good example of the use of these smart tools can be presented also in the city of Volos. Indeed, after the 1955 devastating earthquake, the largely destroyed large part of the city was rebuilt by following an orthogonal layout plan where horizontal (toward the seafront) streets manage both main urban and transit traffic and the perpendicular ones the secondary traffic. In this sense, between the two main horizontal road axes of the city center (see **Figure 11**), the municipality during the SNM study requested to access the possibility to rehabilitate the acoustic environment between the perpendicular street network by means of full or even semi-pedestrianization. L_{den} and L_{night} contours were predicted resulting in that no impact on noise exposure is to be expected for these mitigation measures with most of the building facades in this internal network to be still exposed at L_{den} levels of 75 dB(A) shown in the figure in blue color.

Figure 9. (Left) Attiki Odos selected partial covers, under construction and in use [12].

Figure 10. Location of the four roundabouts already in full operation in Volos, Greece [12].

Figure 11. L_{den} noise action plan maps—impact of the noise mitigation measures (cancel traffic in all perpendicular small streets at Volos city center) (2012) [4].

The smartness of these tools made therefore it possible to evaluate that the noise impact of this small-scale traffic cancellation in the affected inner network would not change the noise exposure of the residential buildings; therefore the noise factor was not a potential evaluation parameter in order to decide such an important measure within the city.

2.4 Smart enough?

The European Directive 2002/49 has established for all Member States a legal and technical framework for managing noise issues in large urban centers and along roads and railways and in the vicinity of airports and industries. The intelligence of these tools lies mainly in the way of measuring and predicting noise and introducing criteria that offer longtime period indices (a whole day/a whole year). These tools have been used for more than 18 years until now and have allowed to have a very precise idea of the environmental conditions in which the inhabitants are exposed to noise. The strategic noise maps are associated with noise action plans accessed by transportation and noise specialists in collaboration with city planners, architects, and policy-makers in order to minimize the impact of environmental noise on the population. The directive gives powerful and intelligent tools to observe the existing situation and its expected development. For example, between 2008 and 2010, Attiki Odos, by monitoring the traffic noise, realized that the noise emissions diminished because the Athenians were reducing the use of their vehicles forced by the economic pressure imposed by governments during the crisis in the country [13].

Noise action plans have to be published and publicly discussed between citizen and policy-makers. These important public meetings are delicate because they support environmental and political disputes that often go beyond the scope of the directive. In several cases residents do not fully understand both measurements and simulation in the strategic noise maps and especially noise simulations adjusted on the facade of their building. In general they consider themselves more exposed and therefore more annoyed compared to the relevant strategic noise maps suggest. There are many reasons for that. The environmental noise is predominant in the city, but it is not based on only one distinct potential source of discomfort. Other sources, such as two-wheeled motorcycles and motopeds, heavy vehicles, amplified music emissions, and neighborhood noise, are in several cases far more important. For these cases, the 24-h measurement and prediction of both the L_{den} and L_{night} indexes as average per a year period do not reflect what a given inhabitant experiences in their everyday life. The political dimension is particularly important, and many municipal councils hesitate to communicate any result because they are afraid to generate more complaints after the publication of both SNM and NAP that must be explained, and public discussion might choose in between the various options available that commits public funding.

In many cases and especially in Members States of Southern Europe, the noise action plans are not always considered as an obligation by the policy-makers and the head of the municipalities. From a legal point of view, it is very difficult to depict clearly the responsible if the objectives of the action plan are not achieved, and this is generated by important bureaucratic obstacles and the local legal framework that do not clearly establish the relevant responsibilities among the different branches of the central and regional governments. The municipalities cannot be considered economically responsible for not having met the objectives of the action plan if there is a lack of necessary funding from the central government especially in period of economic crisis as recently in several Member States of the EU. This is especially true for the municipalities, but this is different in the case of private transportation network operators who are responsible for monitoring environmental parameters of their infrastructure by receiving, accessing, and resolving relevant complaints from local residents. By introducing continuous noise monitoring programs and noise mapping in order to verify compliance of the enforced limit values in order to protect inhabitants from noise exposure by implementing appropriate measurement mitigation and operation measures [5, 10].

3. Environmental noise and soundscape action plans as new smart tools for city development

3.1. Necessary evolution and the smart tools associated with the European Directive 2002/49/EC

Not all the EU Member States have followed the same pace in the implementation of the European Directive. Some published their strategic noise maps online, very fast, immediately after the directive enforcement, because it corresponds a clear political will of the decision-makers; some delayed because they needed the directive to be introduced in their respective national legislation. In some countries, several rounds succeeded one another based on the directive's provision to update the data and the relevant results every 5 years (three rounds until now).

Whatever the case in which the Member States found themselves, after so many years of operation, many thought that the situation could be improved and achieve a homogenous level of completion.

The primary issue discussed and accessed in the relevant EU committees was to establish a more correct and homogenous methodology for calculating and simulating the propagation of the environmental noise sources introduced by the directive. In fact, the method used until 2018 has often been criticized for not being sufficiently precise as regards the emitted noise of different sources and the effects of soil on propagation. Technical improvements have been proposed and adopted by all users by introducing recently the Commission Directive (EU) 2015/996 of 19 May 2015 establishing common noise assessment methods according to Directive 2002/49/EC of the European Parliament and of the Council, introducing the CNOSSOS-EU methodology, to be enforced obligatorily, by all Members States on 31 December 2018. In particular, within this methodology, two-wheeled noise and a new aircraft database were taken into account in the calculations, simulating noise events that in most urban situations are consisting of an acoustic degradation factor, by themselves [8]. The annual average of the L_{den} and L_{day} indices tends to erase the noisy passage of two wheels or a specific aircraft near a receptor (front of a given building) not only in terms of sound energy received but rather on the impact of the average value to express (or not) an annoyance. In the same way also the other modes of transport as the railway have been also introduced in order to better take into account the specifics of each sound source.

Thus, when Greece and Cyprus start implementing the European Directive 2002/49, the country was inspired by other European similar study cases and had also the capacity to move on several open discussions and innovative approaches. Indeed, from 2012, without interruption, the main urban agglomerations of these SE countries were able to publish their results on SNM and NAP regarding the noise environment: Volos, Larissa, Chania, Heraklion, Agrinio, Corfu, Thessaloniki, Athens, Nicosia, Larnaka, and Lemessos. Especially in Greece the operation started with medium-sized agglomerations in Volos and Larissa in 2012, and it led the authors to propose specific adjustments.

Indeed, starting with the city of Volos in central Greece, we have proposed to proceed a little further than the directive's exact specifications and demands.

Firstly, because already extensive measurement monitoring programs were executed [4, 9], it was established that the levels of the directive's noise indices were compatible with the measured ones but relatively low in a general point of view and outside the influence of the main road axes, population is not exposed to high levels as per the national legislation. However, a noise action plan was drafted, including a general plan aiming to preserve the qualities of the sound environment (especially actions to enforce where and when the urban environment is not too noisy). The NAPs are calibrated based on the acoustic monitoring program in the city and have taken also into account a large interview campaign with residents of five selected neighborhoods within the urban agglomeration. The interview campaign, through comprehensive questionnaires performed in local residents, aimed to describe the sounds they hear on a typical week day, to establish the noise sources that they like and those that are uncomfortable, and finally to assess the sources of environmental noise when they perceive them, for example, at home, at work, etc.

3.2. Soundscape issue and inhabitants' perception

Many cities in Europe are undergoing major structural changes and are investing heavily to accommodate more than 70% of the world's population that is projected to be living in urban areas [14] by 2050. Cities are becoming increasingly dense and are forced to implement more and more diversified transport offers. Of course the so-called ecological transport is more numerous, but it is not sure whether they are quieter. Indeed, mass public transport is increasing, solving road traffic problems but not necessarily lowering the noise levels to which people are exposed. At the same time, airplane traffic is exploding and projection gives in this sector. In this context, to create new urban centers and minimize travel, many cities in Europe are trying to build eco-neighborhoods or eco-districts in which all energy dimensions are particularly studied. It is interesting to note that the contribution of the European Directive and its intelligent tools moves from a simple reduction of noise sources to a more qualitative management of the sound environment [15]. The sound dimension is still a dimension of the projects which is not treated as much as that of the energy consumption, but it does not prevent that these questions are now around a global strategy of application of the European Directive. Urban planners, architects, and engineers tried to apply the following principles [16]:

- ❖ Remoteness of dwellings and points of interest from major transport noisy infrastructures
- ❖ Protection of buildings by noise barriers, mounds, and site topography
- ❖ Protection of public spaces and sensitive buildings by using other less sensitive building as a "noise barriers" (parking, commercial spaces, offices, industries)
- ❖ Orientation of the buildings according to the strategic noise maps of noise and potential for apartment openings on calm areas
- ❖ Maximum reduction of the use of the car in these spaces
- ❖ Promotes shared modes of transport and soft and alternative modes
- ❖ Promotes the presence of vegetation, loose soil, and "natural" sound sources But more fundamentally, the Greek experience in the application of the

Directive 2002/49/EC has revealed another dimension which has led the authors to propose specific adjustments. In 2012 (relatively late compared to other Member States and the provisions of the directive), with the experience of applying this directive on the country's main transport infrastructures (roads and airports), the engineering teams, in collaboration with the transportation environmental acoustics and architects, introduced qualitative soundscape analysis tools toward a more efficient assessment and a complete list of recommendations relative to the quality of the sound environment.

They practically note the discrepancy between the values of the relevant noise indices and the common perception of urban sound environments. These elements are all more glaring as the sources of environmental noise are relatively low and much less troublesome. In these medium urban agglomerations, because of their size and their evolution process, residents defend very strongly the identity of their neighborhood, and they describe the sound qualities of these neighborhoods as a very important element in their style of life.

3.3. Toward a smarter tool for urban development: soundscape mapping

The European Directive on noise environment has introduced the possibility for all Member States to develop a specific methodology in order to preserve and protect quite areas. The directive gives several recommendations, and many cities in Europe develop their own guidelines to identify them and protect them. Climate conditions and social behaviors are however quite different between, for example, London and Thessaloniki, so, once again, the smart thing to do was not to decide what it could be good as a max noise level for all involved in a so diverse European Union. Quantitative criteria (L_{den}, L_{night}, $L_{evening}$, L_{day}) had to be completed with more qualitative criteria, and the notion of soundscapes was useful for that: "a soundscape is the acoustic environment as perceived by humans, in context" popularized by Schafer [17] who describes how people like to listen to the sounds and the noises of their environment when they are not annoyed and when they describe qualities of their neighborhood.

In this context, NAPs were completed with soundscape action plans based on the analysis of the relevant quantitative mapping. Many times, specific areas are selected because they are representative of noise and soundscape issues in link with urban development. Several strategies are defined for the protection, the management, and the creation on soundscapes in these areas.

The identification of these zones allows its protection and restoration of those responsible for the development of the agglomeration's urban space (municipalities, architects, urban planners) within the physical city development. Consequently, this quantitative measure does not translate itself all the quality parameters of the acoustic environment of the area. For these reasons we consider it interesting to grow along with the "quantitative" mapping and a "qualitative" mapping of the acoustic environment. To ensure the appropriate assessment tools to city authorities in order for them to act on upgrading the sound identity of the subregion proposed below, the following mapping investigations have been realized.

The city center of Thessaloniki where we applied this methodology is described hereafter based on the following series of diverse layers of mapping [18]:

- ❖ Urban typology map: This map describes mainly the propagation space of sounds and noise and shows on 2D drawings the section of the streets, road, boulevard, and avenues of the studied area (U- or L-shaped roads or open road). This map makes it possible to evaluate the qualities of the urban spaces in which sounds and noise spread (sound space more or less closed, even reverberant or open space) (**Figure 12**).

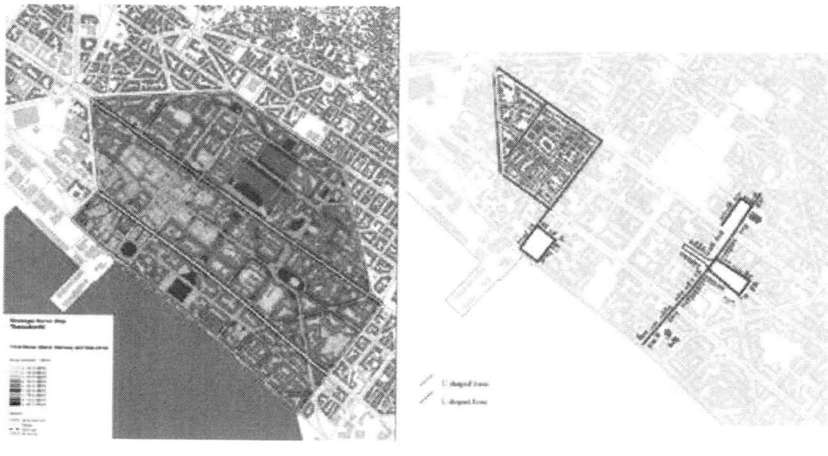

Figure 12. L_{den} **strategic noise map for Thessaloniki city center (left) and the relevant urban typology map of the center area (right).**

- Spatio-acoustic typology map: This map presents the potential acoustic effect that can be created by the urban forms: filers, reverberation, silence islet, cut- ting effect, etc.) (**Figure 13**).
- Map of predominant uses of the public spaces: This map shows the most prevalent uses of public spaces (traffic, shops and shopping areas, services, pedestrians, services, etc.)
- Map of predominant uses of building uses: This map presents the main uses of the buildings of the area studied: residential building, point of interests, shops and shopping areas, services, industrials, crafting, etc. (**Figure 14**).
- Map of sound markers and sound signals of identity's characteristics: This map shows the sounds that characterize a place and that are often quoted by residents (**Figure 15**).
- Soundscape maps: This map is a drawing of different areas where the sound- scape that one can experiment is remarkable and has been described by the majority of people interviewed. It describes the main sound and noise sources heard on site, their relation in intensity and in time, and the way they are interpreted by the residents. This map actually regroups the results of the previous analysis (**Figure 16**).

The creation of these mapping databases in correlation with the quantitative noise measurements allows in-depth analysis of the acoustic qualities and noise characteristics of a given neighborhood while they are not only clarifying the reasons for acoustic quality existence at the neighborhood scale but also annoyance problems. All the previous maps are fully correlated with the relevant noise action plan map of the area produced as per the European Directive guidelines (**Figure 17**).

Figure 13. Spatio-acoustic typology map for Thessaloniki city center (right) urban typology map.

At the same time, they facilitate the decision-making in relation to the urban agglomeration planning (sources, propagation conditions, ground coverings, social organization of the city, etc.). All these overlapping maps consist for the authors a significant improvement of the available tools provided by the directive. These tools make it possible to articulate a complete quantitative and quantitative approach to the urban space and introduce, as well as the physical acoustic one, the results of interviews and all *in situ* observations. In this way, these tools compile users' points of view and produce a more general vision of how the sound environment of a neighborhood is perceived by the inhabitants. He accumulates the opinions and the points of view and allows to evaluate the factors that shape the originality of these places. It does not reduce the potential noise problems that managers have to deal with but, on the contrary, emphasizes their magnitude and characteristics by drafting a noise action plan for the soundscape as well. These plans aim also to reduce the areas exposed to high noise levels, thus not only reducing the noise exposure of residents but also preserving, managing, or even creating new soundscapes.

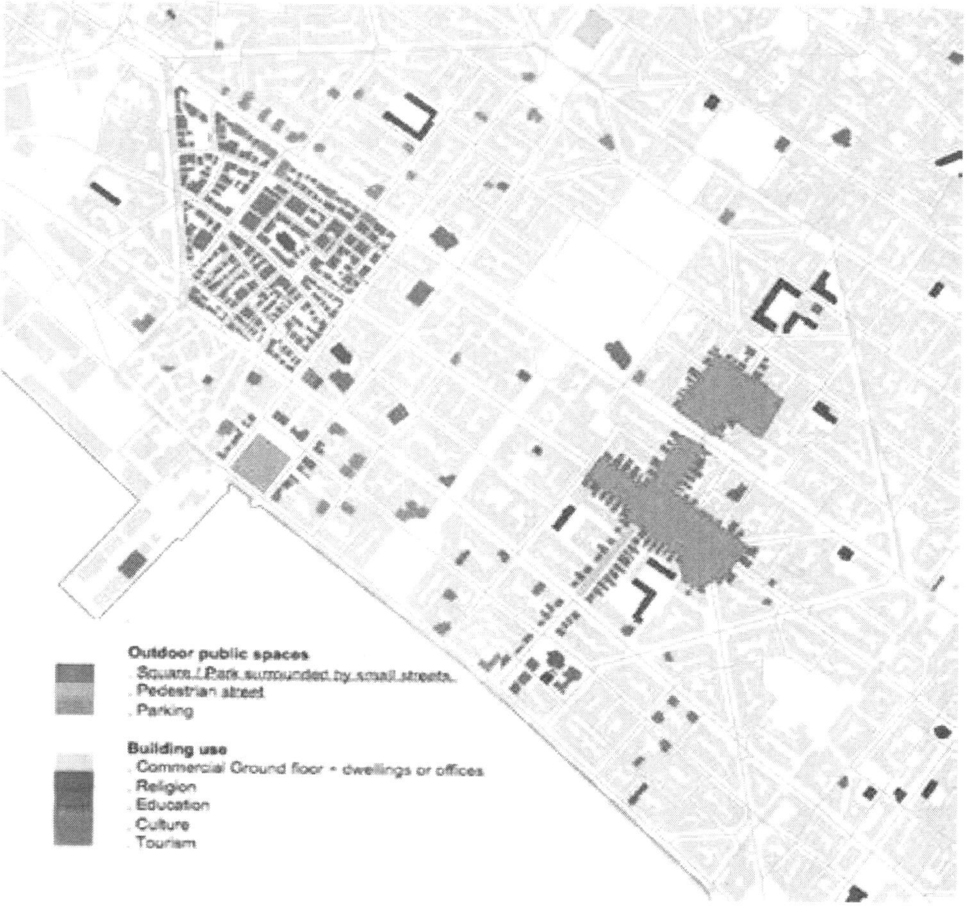

Figure 14. Map of predominant uses of the public spaces and building in Thessaloniki city center.

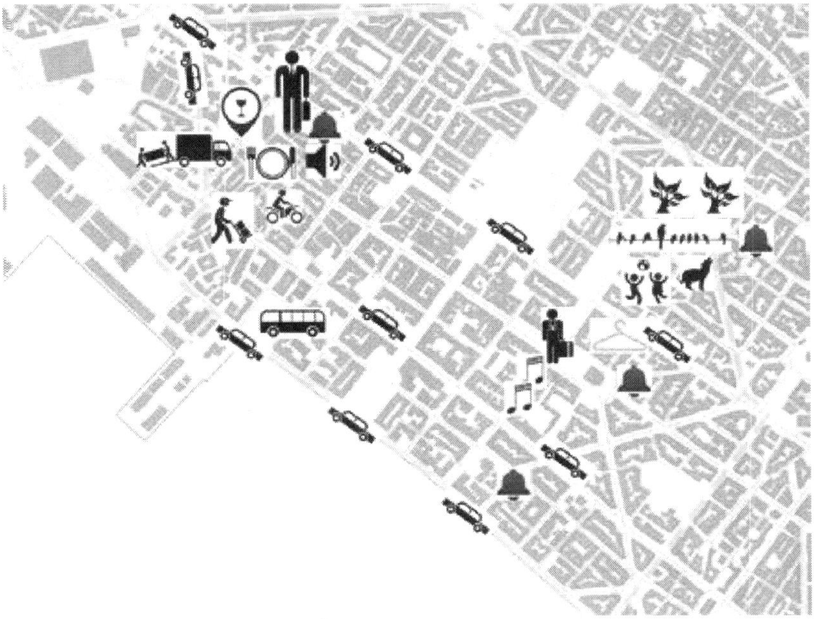

Figure 15. Map of sound markers and sound signals of identity's characteristics of Thessaloniki city center.

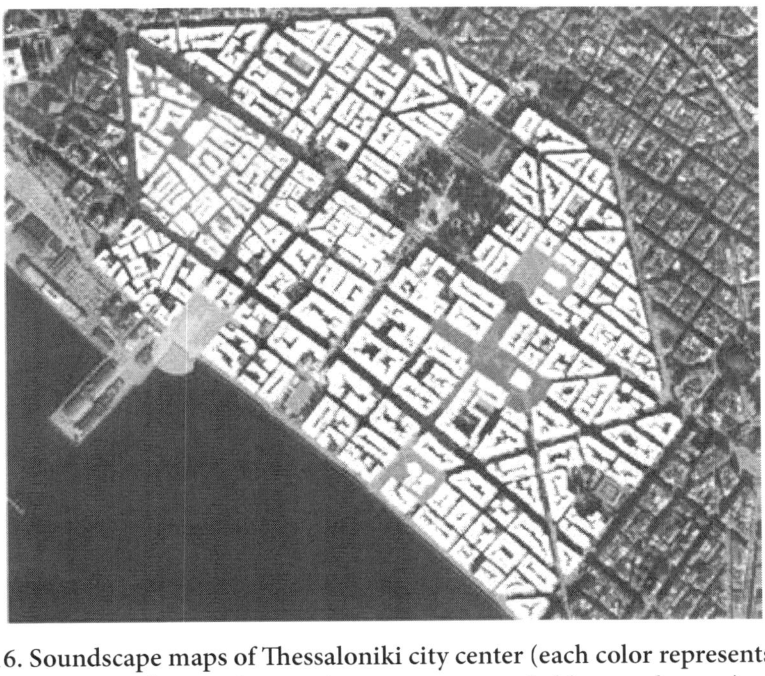

Figure 16. Soundscape maps of Thessaloniki city center (each color represents an area where pass by people experiment some remarkable soundscapes).

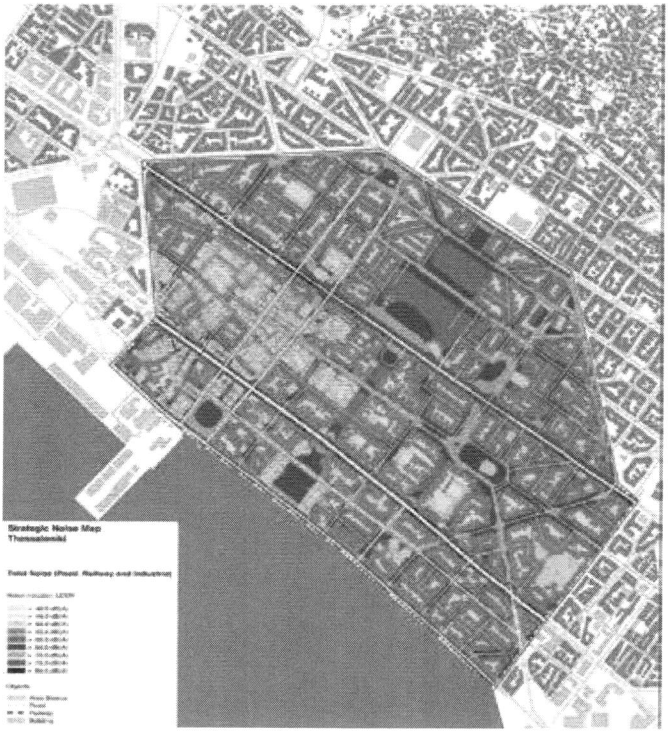

Figure 17. Noise action plan maps for Thessaloniki city center.

Figure 18. (From left to right) Stock Exchange District, Freedom Square, and Agia Sophia Achiropiitou axis localization in Thessaloniki center.

In the example, the historic center of Thessaloniki, which is the subject of numerous renovation projects, the specifics of the action plans can be summarized as follows (**Figure 18**):

Thessaloniki will be equipped shortly with a very modern subway whose main objective will of course be to decongest the urban arteries from the excessive road traffic of today. In this sense, a lot of public space has been the subject of international architectural competitions aiming at their rehabilitation and renovation. In this sense, NAPs, completed by a relevant soundscape action plan (SAP), will allow to introduce a series of development for the city and its neighborhoods:

- ❖ The rehabilitation principle (especially in Stock Exchange District) will have a positive impact on noise exposure. By reducing in general the use of the private car in such environment, the impact will be important because the buildings themselves by their masses and their heights will protect the area from main surrounding circulation axes. By limiting car traffic (streets becoming pedestrian and semi-pedestrian), it will ensure higher importance to the sounds characterizing the recreational and touristic activities' sound signatures (cof- fee and food places, taverns, bars, clubs, live music, shopping).
- ❖ On the north-south axes of Saint Sophia and Achiropiitou (east side of city center), the architectural project selected defends the idea to introduce more (sounds) of nature along the street: water fountains and surfaces, pedestrian areas, benches, etc. will give the opportunity to residents, consumers, and tourists to enjoy the location during the whole week. The light slope from north to south (until the sea level) with relevant urban interventions will help to disconnect the square from the noise traffic impact from Egnatia Avenue, one of the most busy road axes of Thessaloniki. In this case, noise reduction within soundscape creations is expected to manage the main sound ambiances for this district and for the next years.
- ❖ Regarding finally Freedom Square, the challenge was to radically change its architectural image. From the visual aspects, the architectural competition selected a project that will highlight the square. From the acoustic point of view, the challenge is much harder because actually the square is only used by road traffic, parking, and also bus and taxi stations. The natural parameter is highlighted by planting more trees, deciduous and evergreen, and by using on the ground a combination of soil and aged blocks. The project increase also the spaces dedicated to pedestrians by closing the south part of a street.
- ❖ These actions will change the sonic identity of the place if they are fully implemented and then properly maintained (especially regarding the vegetation introduced in the area). It will not be expected to achieve important reduction of the noise exposure from road traffic especially in the sea front, but it will change the space propagation properties and the inhabitant perception. The foreseen interventions is expected to offer several new ways to use this area in an enhanced sound environment, implementing adequate seating possibilities and meeting points, coffee shops, cultural exhibition areas, and with the parking area to be relocated.

4. Conclusions

The tools presented in this chapter can be considered as quite intelligent because they can handle a large amount of data related to environmental noise and the urban soundscape. The mapping features representing these data and their relevant analysis, coupled with the use of detailed geographic information systems, allow to reveal a number of strategies to reduce residents' noise exposure and negative reactions and, above all, to ensure a quality sound environment (soundscape) that characterizes their neighborhood and their city.

Environmental Noise Directives 2002/49/EC and 2015/996/EC need to be implemented along with soundscape analysis in order to propose a more extended and complete noise action plan that considers the urban environment as a whole and not only specific noise sources. In case studies where the environmental noise is very important (as for example, in proximity of major international airports) such study give guidelines to follow for several scenarios and of course including also severe operation measures and even relocation if needed. By embracing a broader framework, acoustic and transportation consultants along with the municipality's officials may develop efficient tools and comprehensive noise action plans that go beyond the simplified issue of noise and offer an expanded view of the situation. The question is therefore not only to specify the tools to develop the city without noise but especially the use of intelligent tools that allow a city to evolve with all its sonorities and soundscapes, toward the noise abatement which is undoubtedly the first preoccupation of the Member States managing adequately all environmental noise dimensions and introducing the proper solutions. An action guide for environmental noise and the soundscape is there- fore a powerful intelligent tool that seeks to manage an environmental problem while keeping what makes the identity (sound) of neighborhoods, all over urban agglomerations in Europe.

The main criticism that can be formulated about this approach lies in the forms of consultation of residents and citizens. Until now, it is often more practical to conduct in situ interviews with the residents of the area. The duration of studies, constrained for economic reasons, does not allow time to "hear the opinion" of everyone. Although the survey techniques used show recurrences in the opinions of interviewees, one could imagine that a system of automatic soundscape perception could be more effective than the method used. Citizen participation through mobile phones for the measuring and the qualification of noise sources and soundscapes has been developed in the recent years and might be used in this purpose. Noise- Capture is described as the scientific tool for environmental noise assessment [19]. The project gives the opportunity to any Android mobile phone to participate in the creation of a strategic noise map. The tool offers the capacity to share the measure- ment and display maps created by all the users, for example, at Vieux Port area in Marseilles [20] (**Figures 19** and **20**).

On both figures above, and at different scales, these tools present a new approach of strategic noise mapping, by indicating noise value and noise source characteristics recorded (noises, soundscapes, etc.). This map created by various independent users depending on the hour of the day, the duration of the measurement, allows an interesting representation of the sound environment as experi- mented by the residents.

These new tools are complementary to the European Directives' provisions and methodological tools, but indeed they are somehow smarter in their ability to massively aggregate noise measurements, predictions, and comments of residents.

Therefore the environmental mapping will introduce new ways of representing complex and dynamic sound phenomena in an urban area ensuring deeper analysis in

Figure 19. Noise map visualization at Vieux Port of Marseilles (NoiseCapture application) [19, 20].

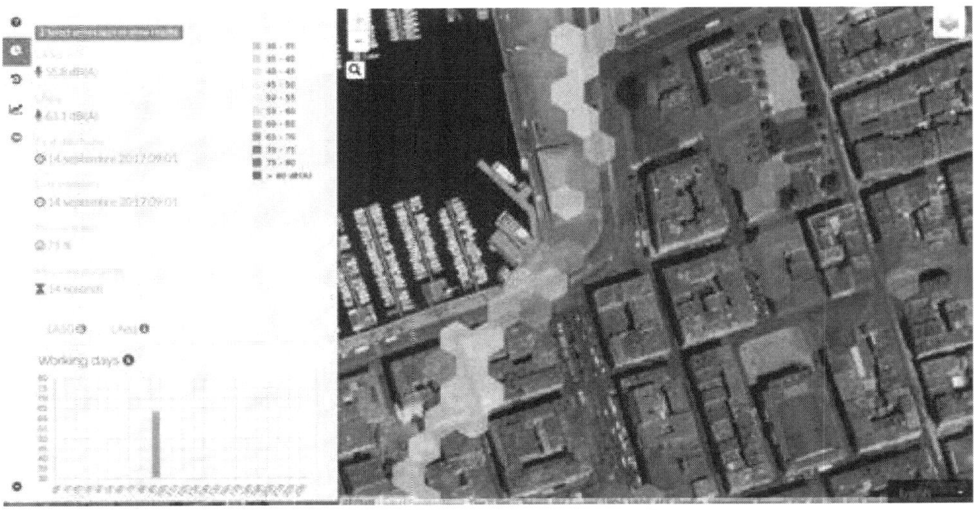

Figure 20. Zoomed in view of Vieux Port of Marseilles (NoiseCapture application) [19, 20].

order to understand and fully access all elements of the soundscape contributing in the formation of the sound identity of neighborhoods and cities. Enriched with all these approaches, there is no doubt that the city will be better equipped by many intelligent tools to proceed in its development by ensuring a sustainable sound environment.

References

1. Directive 2002/49/EC of the European Parliament and of the Council of 25 June 2002 relating to the assessment and management of environmental noise 2019. Available from: https://eur-lex.europa.eu/legal- content/EN/TXT/PDF/?uri=CELEX:32002L0049&from=EN
2. OJ C 337 E, 28.11.2000. p. 251
3. Vogiatzis K. Strategic environmental noise mapping & action plans in Athens ring road (Attiki Odos): Greece. WSEAS Transactions on Environment and Development. 2011;7(10):315-323. ISSN: 1790-5079. Available from: http:// www.wseas.us/e-library/transactions/ environment/2011/54-290.pdf
4. http://www.ypeka.gr/Default. aspx?tabid=936&language=en-US [Accessed: 4 July 2019]
5. Vogiatzis K, Dimitriou D, Gerolymatou G, O'Connor CM, Aristeidis Konstantinidis A. Strategic noise map 2017 of Athens International Airport, Eleftherios Venizelos (traffic data 2016). In: Euronoise 2018, Conference Proceedings; ISSN: 2226-5147; 2018
6. Vogiatzis KE, Kouroussis G. Environmental road transportation noise monitoring: a necessary tool for the protection & rehabilitation of the urban acoustic environment. In: Proceedings of the 24th International Congress on Sound and Vibration, London (UK); 23-27 July 2017
7. European Environment Agency portal. Noise Observation and Information Service for Europe [Internet]. Available from: http://noise. eea.europa.eu [Accessed: 21 June 2019]
8. Commission Directive (EU) 2015/996 of 19 May 2015 establishing common noise assessment methods according to Directive 2002/49/EC of the European Parliament and of the Council. Available from: https:// eur-lex.europa.eu/eli/dir/2015/996/oj [Accessed: 4 July 2019]
9. Vogiatzis K, Remy N. Changing the urban sound environment in Greece: A guide based on selected case studies of strategic noise maps (SNM) and noise action plans (NAP) in medium and large urban areas. Environments. 2018;5:64; [Accessed: 21 June 2019]. DOI: 10.3390/ environments5060064
10. Vogiatzis K, Remy N. Strategic noise mapping of Herakleion: The aircraft noise impact as a factor of the Int. airport relocation. Noise Mapping. 2014;1(1):15-31, [Accessed: 21 June 2019]. DOI: 10.2478/noise-2014-0003
11. Vogiatzis K, Zafiropoulou V. Implementation of noise barriers in Attiki Odos Motorway based on the relevant strategic noise mapping and noise action plan. In: Euronoise 2018: Conference Proceedings; EAA—HELINA; ISSN: 2226-5147; pp. 1175-1180
12. www.google.com/maps/place/ Athènes [Accessed: 21 June 2019]

13. Vogiatzis K, Kopelias P. Benefits & limitations toward a sustainable urban environment during the years of economic recession. International Journal of Sustainable Development and Planning. 2015;10(5):701-712

14. OECD Environmental Outlook to 2050: The Consequences of Inaction— Key Facts and Figures [Internet]. Available from: https://www.oecd.org/ env/indicators-modelling-outlooks/oec denvironmentaloutlookto2050theconse quencesofinaction-keyfactsandfigures. htm [Accessed: 21 June 2019]

15. Ecoquartiers et environnement sonore: De la lutte contre le bruit à la conception d'une ambiance sonore confortable [Internet]. Available from: http://www.bruit.fr/ nos-services-aux-acteurs-du-bruit/ colloques-et-journees-techniques/ colloques-et-journees-techniques- les-archives/ecoquartiers-et- environnement-sonore-de-la-lutte-contre-le-bruit-a-la-conception-dune-a. html [Accessed: 21 June 2019]

16. Rémy N, Marchal T, Chelkoff G, Pirhosseinloo-Amini H, Bardyn J-L, et al. Esquis'Sons ! Sketching soundscapes by using parametric tools: Application to the design of balconies, loggias, terraces and corridors of building facades. In: Euronoise 2018; May 2018; Hersonissos, Crete, Greece. Available from: http://www.euronoise2018. eu/ docs/papers/387_Euronoise2018.pdf [Accessed: 21 June 2019]

17. Schafer RM. The Tuning of the World. New York: Random House Inc.; 1977. ISBN: 0-394-40966-3

18. Coelho B et al. Urban sound planning in the city centre of Thessaloniki. In: ICSV22; 12-16 July 2015; Florence: Italy; pp. 1806-1814

19. Noise Capture project: Noise-Planet Scientific Tools for Environmental Noise Assessment [Internet]. Available from: http://noise-planet.org/project.html [Accessed: 21 June 2019]

20. Bocher E et al. OnoM@p: A Spatial Data Infrastructure Dedicated to Noise Monitoring Based on Volunteers Measurements [Internet]. 2016. Available from: https://peerj. com/ preprints/2273/ [Accessed: 21 June 2019]

Section 4
Smart Urban Development and Mobility

Chapter-9

Understanding Urban Mobility and Pedestrian Movement

1. Introduction

Currently, the field of urban mobility modeling is experiencing a surge of activ- ity due, in part, to renewed interest in crowd management (including evacuations due to natural and man-made disasters) but also influenced by increased efforts to reduce CO_2 emissions through optimization of urban networks for both traffic and pedestrian purposes [1, 2]. Urban sprawl is a recognized phenomenon for growing cities, and tools, such as urban growth models, have proven valuable for planners and decision-makers in identifying challenges and potential environmental impacts [3]. Expansion of the built environment to meet population demand implies extended daily commutes as well as loss of other land function and is recognized as a critical challenge in global change, not only in countries experiencing explosive industrialization but worldwide [4–9]. Growth in population size of many major cities presents complex logistics in meeting demands for increased numbers of daily commuters and alternative transport modalities. In the UK, for example, the 11 most populous cities since 2015 are to be found in Scotland, (Glasgow and Edinburgh); the conurbations of Northeast England, the West Midlands, and South and West Yorkshire (adjacent to the cities of Greater Manchester and Liverpool); Bristol and Cardiff in the southwest; and, of course, Greater London [10]. Between mid-2011 and mid-2015, Greater London's population grew by 5.7% to around 8.67 million, compared to that of other city regions (2.3%) and to the average growth (2.7%) for the country as a whole.

Under pressures of increased population growth, short-term crises and long- term policies, city layouts and infrastructure constantly adapt to meet needs, but the many factors involved render solutions for high volume passenger movement far from trivial. Awareness of the consequences of unrestricted urban sprawl has motivated legislation and a global move toward environmental sustainability over several decades, but change is slow [11]. The performance and modalities of transportation networks have attracted considerable attention, fueled mainly by efforts to reduce road congestion and harmful emissions. For example, Transport for London (TfL) (created in 2000) manages the capital's principal road networks, the underground system and its extension, the Docklands Light Railway, and TfL Rail (responsible in conjunction with the Department of Transport for commissioning crossrail, designed to improve East–West transit). While the TfL budget (~10 billion sterling in recent years) demonstrates major commitment to maintenance and new development, its business scorecard also emphasizes the need for a system accessible to all, the "greening" of the city streets and the health benefits for Londoners "traveling actively" [12]. Accommodating positive lifestyle choices such as walking or cycling commutes, as well as decreasing the CO_2 burden from road traffic, has served also to shift more attention toward the pedestrian's city experience. In consequence, this chapter also explores the implications for "traveling actively," and safely, in London.

From the pedestrian viewpoint, the need for green spaces in city planning has long been recognized [13], but factors for *active travel* remain complex. Digital street mapping and mobile technology have improved familiarity and navigation within the urban environment, but, while novel modes of individual transport (such as smart urban bicycles and electric scooters) reduce the emission burden, road usage is increasingly multifaceted. Inevitably therefore, strategic emergency management is complicated by the challenge of prompt multimodal evacuation of dense urban areas [14]. In discussing plausible modeling approaches which capture principal factors influencing rapid transit to daily destinations (such as schools and offices), as well as leisure trips to parks and entertainment venues, consideration is given not only to throughput but also efficient evacuation from these high-density locations. The focus, specifically, is on the flexibility which agent-based model- ing brings to representing pedestrian behavior. The paradigm permits individual actions, awareness of the environment, and pedestrian group movements to be modeled simultaneously on real urban networks.

Pedestrians are distinguished by a number of key features, such as personal choice, variable dynamics, and vulnerability. Debatably, they have the least predict- able behavior patterns, although it has been shown that crowded venues restrict optimal choice [15–18]. Specifically, it has long been demonstrated that pedestrians can move freely *only* when pedestrian densities are small [15]. Designing urban infrastructure in order to increase pedestrian activity, therefore, has to balance often conflicting requirements of personal characteristics (such as walking speed), against considerations of safety. The problem is that parameter space is greatly expanded by variation in pedestrian profiles; for example, age, speed, knowledge of the environment, individual or group transit, entrance and exit point to the net- work, time of day, and occupation density (among other factors) all affect efficient transit as well as the logistics of congestion and evacuation. Variable dynamics

Figure 1. Aerial views of Singapore (left) and Zurich (right) urban layouts. Both cities consistently rank in the top 10 in the world for urban layout and mobility [20, 28, 29].

can be illustrated by examples of walking patterns for an average shopper, which are markedly distinct from pedestrians in a business district. Similarly, an elderly person typically moves differently to a young one, as does a native to a tourist and so on. Even within a particular scene, e.g., a shopping district, logistics are different for the successfully laden pedestrian and those still browsing [19].

As a consequence of this diversity, shaping sustainable city infrastructure relies on understanding pedestrian movement patterns and the environmental and behavioral reasons that guide them, together with provision of suitable public transportation options at key locations. Cities with strong track record in infra- structural design for mobility include Singapore and Zurich (**Figure 1**). Arguably due to large budgets, it has been shown that quality and safety of urban infrastructure do not relate solely to wealth, as good planning practices are vital [20]. Looking ahead, GPS-enabled mobile apps are likely to shape pedestrian behavior trends further, with awareness of urban layout (such as important intersections, walking routes, street signs, and transport alternatives), less reliant on physical observation than 'in-app' street map layouts, together with walking time estimates based on the historical consumer mix [21].

Investing resources in sustainable city planning is not for the fainthearted. Burgeoning demand for access and choice continues to threaten limits for air quality, noise, energy consumption, and biodiversity. The last hundred years has seen urban population growth concentrated on less than 3% of the world's surface but with the corresponding environmental footprint disproportionately impacting climate: currently, 75% of greenhouse gas emissions can be attributed to cities with ecological effects many times larger than the actual urban area occupied [22]. Socioeconomic implications, such as health and well-being, are also a cause for concern: in France and elsewhere, urban mobility plans are now a required component of the urban planning process for the future [23], while global city initiatives, such as the 10 Aalborg Commitments [24], attempt to define basic guidelines for sustainable development.

2. Overview of modeling approaches

Within the broader agenda of sustainable urban planning, computer modeling has gained increased popularity as a versatile tool. The ability to explore *in silico* the nature and effect of change can facilitate the planning process, providing insight on the parameters, key dependencies and potential pitfalls, as well as complementing pilot schemes.

Emergency evacuation typically follows natural disasters, terrorist attacks on transport networks or at major events, as well as other causes of injury or where crowd dynamics destabilize [25]. The so-called climatic "extreme events" have markedly increased over the last decade, with ever-more severe consequences [26]. Increased frequency of such events, together with increased population density (mainly concentrated in urban areas and regions experiencing rapid urbanization, such as Asia) [26], has led to some of the largest losses of infrastructure in recent history. Besides highlighting the need for preemptive action and resilient infrastructure, extreme event prediction is widely employed to mitigate the human cost and employ successful evacuation strategies (as in the very recent example of Cyclone Fani's landfall in India and Bangladesh (2019) where more than 2.8 million people were evacuated ahead of the storm) [27].

Approaches to modeling crowd behavior, pedestrian flows, and evacuation methods are varied and range from studies looking at flows of people as a paradigm [30–32] to the analysis of individual behavior patterns [33–36]. Early work aimed to describe pedestrian motion through physical model types including fluid dynamic and social forces, based on Newtonian mechanics [37]. Pedestrian motion can be described, for example, using a sum of different force vectors—namely, attractive, repulsive, driving, and fluctuating. However, the downside of these models is their reliance on sophisticated mathematical expressions that become intractable on expansion for newly discovered parameters and behaviors. Further, individual movement is represented as a superposition of pedestrian interactions, not only nontrivial to solve but often opaque to interpretation [38].

Key features to be incorporated are the agenda of the individual (purpose of journey) as well as interaction with the built and demographic environment—road traffic, urban layout, and crowd size. Two elements present particular difficulty.

Pedestrians do not always follow simple logic or "stimulus-and-response"-based behavior and, unlike other road users (such as motorized vehicles or bicycles) do not need to, and indeed do not, follow preset movement lines. This freedom in choice and execution of movement means that any model must allow for randomness, treating individual behavior as unique to some extent.

2.1. Pedestrian movement

Two main model types can be distinguished for pedestrian interactions, namely, those for route choice and road crossing behavior, respectively. The former category is concerned with optimizing route layouts to achieve shortest travel times between origin and destination under various constraints, such as emergency road closures or congested pathways: investigations of crowd behavior and evacuation dynamics mainly utilize these scenarios, e.g., [39]. In contrast, road crossing models focus on pedestrian decision-making and the nature of interactions on road crossings: here key elements include aspects such as crossing gap (gap acceptance theory) and the use and location of the crossing itself (utility theory), e.g., [40].

Further categorization is possible by model scale, usually denoted micro-scopic or macroscopic. Macroscopic models are often route choice ones and are underpinned by the mathematics of fluid mechanics and queueing theory. Earlier examples include optimization of pedestrian network topologies [41] based on pedestrian queueing networks, representing crowds as single, flowable entities [42] and resolution of bottlenecks by disaggregating upstream and downstream flows around the point of congestion [30]. More recent work includes formulating pedestrian flows as a family of measures and flow maps [43] and vision-based models [44]. Microscopic models currently account for the majority of pedestrian movement studies [45], with some of the first models in this space based on the cellular automata (CA) paradigm [46]. In CA, the environment and street layouts are represented as matrices of cells with individual pedestrians being able to move from cell to cell by discrete steps in a given model iteration. Update between iterations is performed by applying a matrix of cell state translation rules (the transition matrix) to model successive movements (**Figure 2**). Historically, CA models were used to describe various pedestrian movement scenarios in both route choice and pedestrian crossing categories, from bidirectional pedestrian flows on footpaths [33] to interactions of pedestrians with the urban layout [47].

Increase in computing power over the last decade has seen expansion of the CA paradigm with next-generation simulations for pedestrians based on multiple

agents. These multi-agent or agent-based models (ABM) achieve microscopic levels of simulation, based on artificial intelligence concepts [45]. In agent-based systems, pedestrians are modeled as fully autonomous entities with cognitive and behavioral learning characteristics. Early applications included analysis of global movement patterns [50] and impact of pedestrian space allocation during movement [34].

Recent examples include [48, 49] where the former considers interactions of pedestrian agents in counterflow situations and the latter employs ABM to simulate different categories of pedestrian behavior at congestion points in a large city lay- out. The ABM approach, combined with the processing power of large computing clusters, enables effects of individual human choice within precise urban geom- etries to be modeled realistically. The practical potential for the future of city design and provision is considerable (e.g., smart city initiatives—such as [51]).

2.2. Evacuation dynamics

In modeling disaster scenarios, normal pedestrian movement simulation does not apply. Evacuation of metropolitan areas requires rapid crowd dispersion by safe routes to non-hazard zones at short notice. In terms of large-scale natural disasters such as cyclones, circumstances are even more extreme in terms of volume of people movement and area affected; for example, a few million persons might need to be moved to safety from an area of 160 square kilometers [27, 52]. Evacuation models again, therefore, have a clear division by

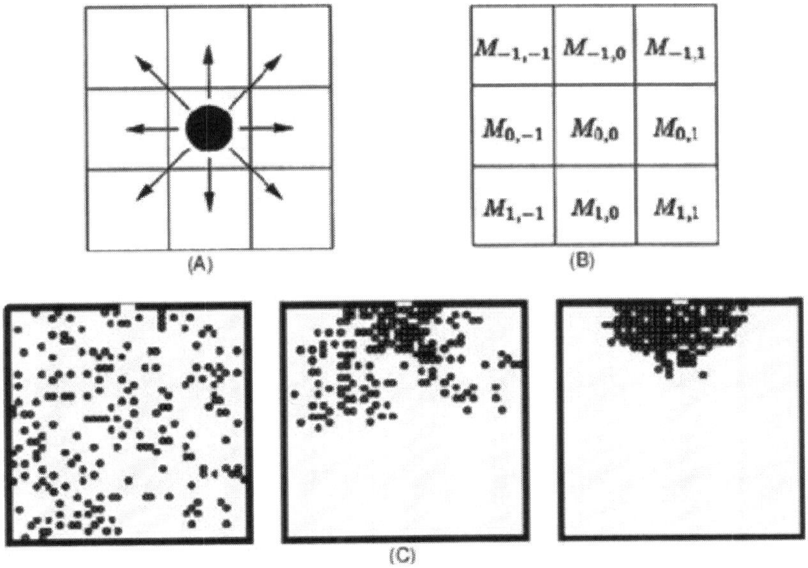

Figure 2. An example of a cellular automata model with transition matrix [54]. (A) A particle (individual) with possible transitions, (B) Matrix of transition probabilities, and (C) Simulation of pedestrians leaving room with single door.

scale, based on the area impacted: small-scale evacuations may involve isolated locations, such as rooms, buildings, and stadia, while large-scale evacuations can include anything from suburban and urban metropolitan areas (with high population density) to tracts of land with different population densities [53].

Microscopic models for building evacuation have been around for some time [54]. A useful categorization is provided by the US National Institute for Standards and Technology (NIST) [55], based on orientation, building type applicability, size of grid, user-perspective, type of behavior, and type of movement. Of particular interest in the NIST nomenclature is the classification of models into behavioral and movement types. Behavioral models simulate action-taking by pedestrians, depending on the specific emergency circumstances, while movement models concentrate on evacuation flows. Models, which incorporate both individual action and evacuation strategies, are classified as mixed.

Further subdivision is possible according to the nature of the behavior exhibited.

Thus, implicit behavior models, conditional behavior models, models utilizing artificial intelligence, and probabilistic models have all been proposed, of which the first are the simplest. The behavioral response of individual pedestrians is built into movement patterns or response delays but is not modeled explicitly as a *conscious choice* [56]. Conditional approaches follow an "if-then" rule pattern—evacuee behavior is modeled as a response to structural characteristics or structural changes in the surrounding environment [57]. All models simulate individual pedestrians through modeling the human intelligence aspect of their behavior directly (as opposed to indirectly via movement parameters as for other model types) [58]. Probabilistic models assign behavior probabilities to individual groups permitting random outcomes for each model run, with statistics analyzed after repeated runs. Compared with AI models, parameterization can be based on summary data for real disaster events [49]. Agent-based models (ABM) (combining both AI and probabilistic approaches) thus offer considerable strengths (discussed in more detail below).

2.3. Model choice

Choice of the right model does not always involve the more complex or even the most realistic since complexity requires a large set of parameters, for which empirical estimates are often unavailable (e.g., profiles of people in a given evacuation context), so simplicity can be an advantage. Moreover, the mode of evacuation can be a critical determinant (applicable almost exclusively to macroscopic models). For large-scale evacuations, the majority of research to date has assumed vehicular transport (predominantly car-based) movement [14].

However, this is sometimes neither practical nor possible and can, on occasion, lead to further escalation of disaster situations by contributing to congestion [59]. In other cases, no such transport option is available, and/or existing public transport cannot be used in the immediate vicinity [60]. In 2005, for example, Hurricane Katrina left 80% of New Orleans in the US state of Mississippi flooded, with some parts under nearly 5 m. of water [61]. In consequence, investigation of exclusively pedestrian-based evacuations in circumstances where utilizing of usual transport modes is not an option (e.g., earthquake disasters or floods) is gaining prominence.

In this context, **Figure 3** illustrates New Orleans (map taken from [62] with vehicular evacuation routes shown in green and population densities in orange). Implications for loss of access to routes for car transport are clear.

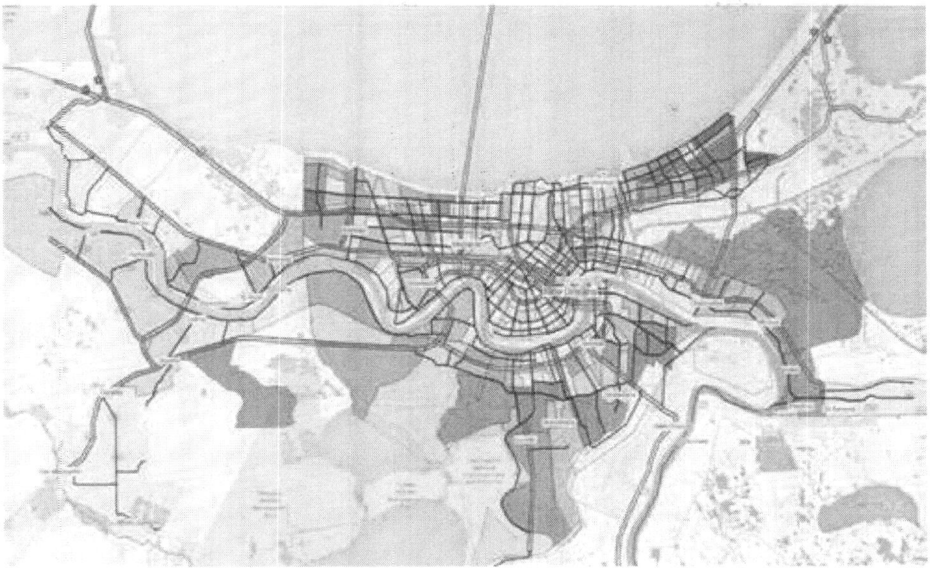

Figure 3. Map of New Orleans showing the sectors of an evacuation plan obtained via optimized modeling. Blue lines indicate secondary roads used in evacuation routes. Green lines indicate roads used as one-way contraflow evacuation routes. The orange shading indicates population density, with darker shading indicating greater density [62].

Well-established early traffic simulation models, such as PARAMICS, VISSIM, and CORSIM [63–65], have recently become popular also for emergency evacuation scenarios, using adjusted parameter values, e.g., acceleration of vehicles and reaction time, which differ in disaster situations [66, 67]. However, other transport options, such as the rail system (arguably an effective evacuation mechanism due to larger capacity), have not been extensively modeled (although included in existing urban evacuation plans (e.g., for Chicago [68])).

Clearly, however, major disruption to (or congestion of) available transport networks, combined with the high population density in urban areas, means that evacuation on foot provides a vital mode of escape. Pedestrian evacuation models of this type have only recently begun to feature in the development of city evacuation plans, while adaptation of existing evacuation model tools is again necessary to accommodate features involved [45].

Increasingly important, however, in modeling both urban mobility and evacuation scenarios are new technology tools, such as volunteered geographic information (VGI) systems. VGI systems allow for collection and dissemination of global urban data, based on user-generated content and peer-review, and thus allow creation and curation of geographical datasets that would otherwise be too cost-prohibitive to assemble for individual research purposes. A good example of a VGI system is OpenStreetMap (OSM), an open project with the purpose of creating non-proprietary geographical maps of the world [69]. Led by the OpenStreetMap Foundation, its stated goal is to encourage the development and distribution of free geospatial data for anyone to use and share. Particularly attractive is its fine-grained coordinate layout and geographic metadata associated with each map element. OSM maps provide a good backdrop on which to develop both CA and AB model types.

These can incorporate both quantitative (e.g., street lengths and lane numbers) and qualitative (street types, nearby amenities) map data to accurately simulate grid "cells" (in the CA type) or free-flowing pedestrian environments (for ABM).

3. Pedestrian behavior

As urban environments expand, routine travel to work or other destinations typically takes longer and can be increasingly affected by congestion and delays for both public and private transport modes. Alternative lifestyle choices such as walking and cycling can prove both healthy and efficient but are also subject to constraints of the built environment and demographics. Although pedestrian behavior has been studied for more than several decades [70], predominantly

with respect to self-organization patterns and interaction of pedestrian flows [20], additional parametrization has become possible relatively recently due to expansion in computing power. In consequence, the questions addressed have become more complex and more relevant for both normal movement and for emergency scenarios. Examples cited include the use of models to analyze evacuation patterns from enclosed spaces (such as buildings, underground stations, and other public venues) [27, 71, 72], to address large-scale problems in morphological urban structure as well as to understand cognitive behavior in the context of disasters (such as hurricanes and terrorist attacks among others) [38, 73].

3.1. Groups or individuals

Addressing self-organization [15, 32], some studies report that, rather than wholly random or individual movement, interactions inside and between groups lead to formation of typical walking patterns. Distinction exists between travel as a single individual or within a group, however, so that while pedestrian behavior is diverse, with each individual permitted flexible options for movement through crowds or definition of "optimal" route, such groups or crowd pressure act as limiting factors to free choice [74]. Equally, knowledge of the built environment and configuration of the urban street network augments visual perception and cognitive understanding of spatial complexity to determine route choice and understanding of the way in which directional change complements distance [75, 76]. Consequently, while motorized (and non-motorized) road-using vehicles are constrained by traffic rules, signalization, and street orientation, pedestrian flows are subject to fewer fixed rules, exhibiting greater randomness at every time point during free movement

[45] but subject to continuous real-time reassessment and rapid adaptation of route choice under congestion. **Figure 4a** and **b** illustrates some of the flexibility of choice available to the pedestrian under his or her perception of advantage to be gained during urban travel.

The figures serve to highlight those properties which strongly motivate bottom- up modeling of pedestrian movement; the agent basis provides a flexible tool for analysis of complex social behavior [78], with agents *actively aware* of their environment (traffic, adjacent pedestrians, and the street network).

3.2. Real urban networks

Perceptions of the network also depend, however, on how well this can be represented, and the importance of VGI (noted above) has led to considerable model refinement. For example, in [49] the authors introduced a discrete, behavior- driven space–time framework, allowing pedestrian movement to be modeled on a real urban network. The main focus is on exploring the potential of the approach through example scenarios and investigation of simple hypotheses of pattern evolution. The research considered pedestrian movement originating from three main "cognitive features" [76, 79]: (i) walking strategy, (ii) spatial awareness, and (iii) knowledge of the urban grid. **Figure 5** shows emergence of flow patterns originating from such features in a hypothetical peak commute hour scenario for several hotspots in the City of London's financial district.

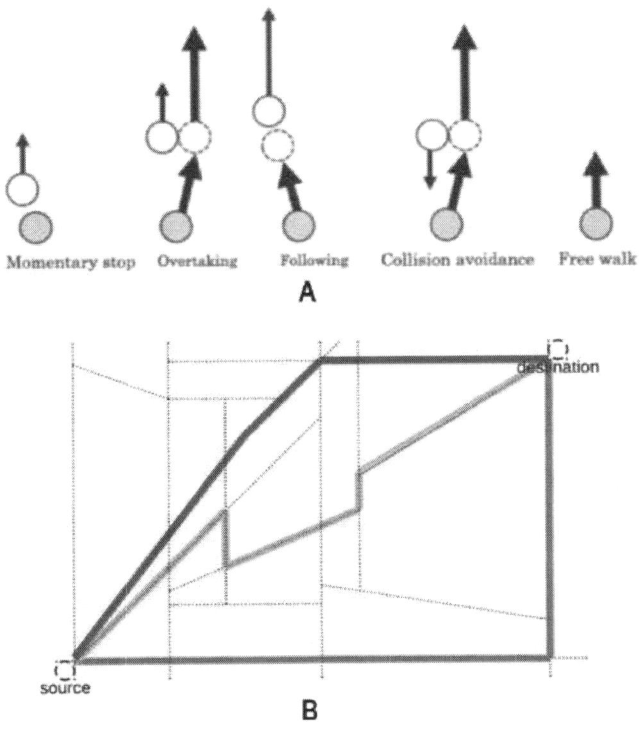

Figure 4. (A) Examples of different pedestrian behavior depending on interaction with other pedestrians during street crossing [77]. (B) Examples of different pedestrian behavior depending on shortest route perception; green, shortest distance is the least actual cost path; red, least angle change requires pedestrians to course correct their path toward "most likely" turns; blue, least turns put the highest "cost" of the route into actual turns needed to reach the destination.

Unfortunately, normal cognitive behavior patterns do not apply in emergen- cies, and route choice during a disaster scenario involves elements that are not present during regular commutes. Among others, these include decision-making under pressure, limited visibility, unclear evacuation routes, and dependency on others in the same group (and in authority) to indicate optimum or safe direction. Additionally, crowd dynamics can change rapidly. It has been shown that crowd turbulence restricts movement at extreme densities (a phenomenon observed during recent crowd disasters) [38] and also modeled by [80, 81]. Thus self- organizing behaviors, designed to optimize motion on the urban network under normal conditions, break down at high crowd densities and for bottlenecks that occur during large evacuation scenarios [38]. Simple patterns, such as formation of unidirectional pedestrian flows in bidirectional traffic, disappear and are replaced with other collective patterns like long-range collisions and stop-and-go waves that lead to serious participant injuries during mass events. In an attempt to understand the forces and factors involved, recent studies have considered merging behavior of pedestrians under different scenarios as well as models for collision avoidance [32, 82, 83].

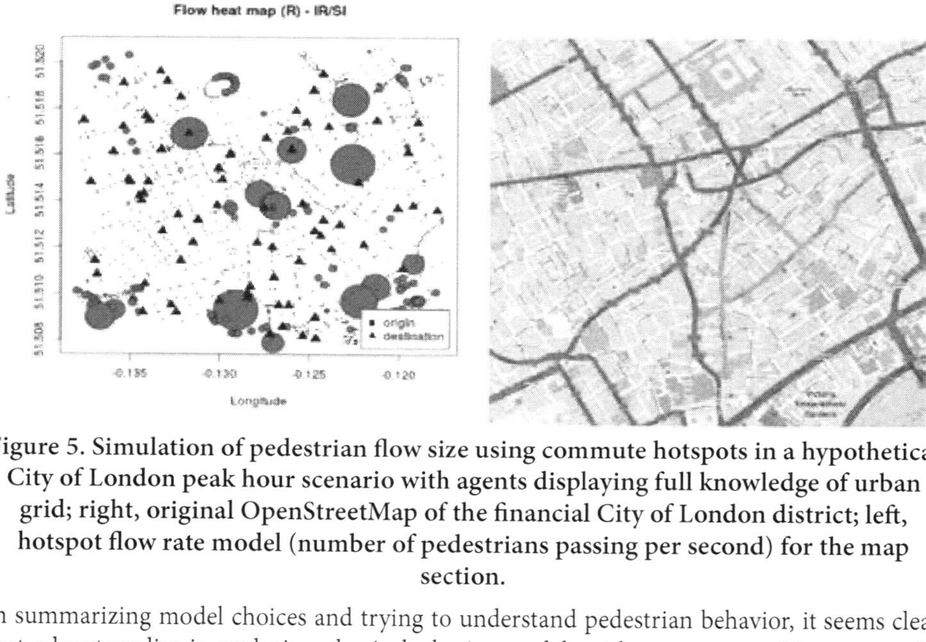

Figure 5. Simulation of pedestrian flow size using commute hotspots in a hypothetical City of London peak hour scenario with agents displaying full knowledge of urban grid; right, original OpenStreetMap of the financial City of London district; left, hotspot flow rate model (number of pedestrians passing per second) for the map section.

In summarizing model choices and trying to understand pedestrian behavior, it seems clear that advantage lies in replacing classical physics models with a more cognitive approach, tailored to single-person (agent) granularity. In evacuation scenarios, in particular, behavior is based on the concept of heuristics, namely, quick and simple cognitive processes that tend to pare down visual perception of the world and optimize for speed (a crucial aspect of emergency decision-making). Agent- based models also permit simulation of well-known "grouping" behavior during such scenes, including cohesive bounds and "herding," where groups of individuals decide to communicate, act, and stay together as a group. These fine-grained clustering aspects of behavior are not well-captured by physical approximation [84].

4. Agent-based modeling

4.1. Advantages and scope

Building on earlier discrete methods (such as cellular automata), agent-based modeling (ABM) has gained considerable popularity for the representation of individual pedestrian interactions. The approach has several key advantages, the most important being the expressive and intuitive nature of the modeling language, its suitability to high-performance execution environments, adaptability to inclusion of heterogeneous behavior, and incorporation of stochasticity [85, 86]. The origins of application of ABM to pedestrian modeling lie in simulations of social behavior and decision-making, introduced in detail in [87]. From early models, where agents of two distinct types populated a simple grid [88], use has expanded to representation of complex real-world situations and social behavior involving millions of entities (e.g., TRANSIMS [89]).

The modeling strengths of the agent-based approach for pedestrian behavior are wide-ranging. Characteristics of individual pedestrians can be defined, including estimates of their spatial awareness using cognition precepts, combining with preferential choices determined for different social groups. ABM can be used to investigate behavior patterns that incorporate rules of movement along pedestrian routes as well as intermediate decision and conflict points.

Dynamic volunteered geographic information system data (such as that from the OpenStreetMap platform) can be utilized, permitting analysis of arbitrary city networks and comparison of the effect of grid structure and amenity distribution. Interaction of multiple social groups can also be investigated, for example, those consisting of pedestrians who have "directed" (e.g., point-to-point) patterns as opposed to those progressing at "leisure" (with patterns that are more random and less easily graphed). Such features offer the potential for these models to explore urban flows and congestion and the way in which changes in network morphology affect route choice. Equally, characteristics of the urban networks in responding to changing demand can also be modeled as well as disruptions impacting individual agent paths and travel times. ABM also compares well with statistical prediction techniques for pedestrian flows that have gained popularity in recent years, such as multiple regression analysis (MRA) [90]. This type of analysis relies on known parameters such as length of pedestrian routes and visual connectivity between points to estimate e.g. throughput numbers per given unit area [91]. While useful for estimating and understanding aggregate numbers representing pedestrian flow data, difficulties arise in accounting for aspects such as urban network architecture and layout [92]. Although agent-based models cannot access real pedestrian movement data on a large-scale urban level to model flows through individual streets, known information about individual pedestrian behavior does enable fine-grained implementation to explore different mobility scenarios at individual street level (within the large city model) as well as stochastic approximation for areas with sparse data.

4.2. Visualization of pedestrian behavior on urban networks

In simulating crowd and group dynamics, ABM enables exploration of force effects at different crowd densities by using discrete grid cells with assigned *force vectors* [93] and demonstration of local patterns for random pedestrian walks, utilizing aspects of both micro- and macro-simulations [94].

In **Figure 6**, an extract from OpenStreetMap shows a section of central London's financial district for which the agent-based model has been used to simulate dif- ferent types of progression, i.e., point-to-point or directional walking (that might relate to a commute) vs. the more random progression (associated perhaps with tourist sight-seeing). The implications for density and dispersion of occupation

are indicated by the coloring. In the first part of the figure, clear preferred routes are the most congested and are colored red, the next preferred yellow, and so on. Hotspots are clearly identified. In the second part of the figure, clustering occurs at "sights" rather than along routes, but hotspots are often offset in terms of access.

Clearly these scenarios represent different challenges in the case of closures or evacuation requirements. Shown specifically here are entrance points to alternative transport modes (black squares and red triangles, respectively), such as the under- ground. In some scenarios these may of course be unavailable or closed down in the immediate emergency zone.

In [49] the authors show how a general agent-based model combined with VGI data can be utilized to describe a wide variety of pedestrian behaviors covering both emergency and non-emergency situations. ABMs perform well in modeling individual pedestrian behavior as *generic state machines*. For every pedestrian we can specify a generic decision-transition-waiting flow. Individual states can then be further broken down to simulate fine-grained psychological or perceptual aspects of individuals. In the example of non-emergency behavior (i.e., daily commute, travel, leisure), we can break down the decision state to

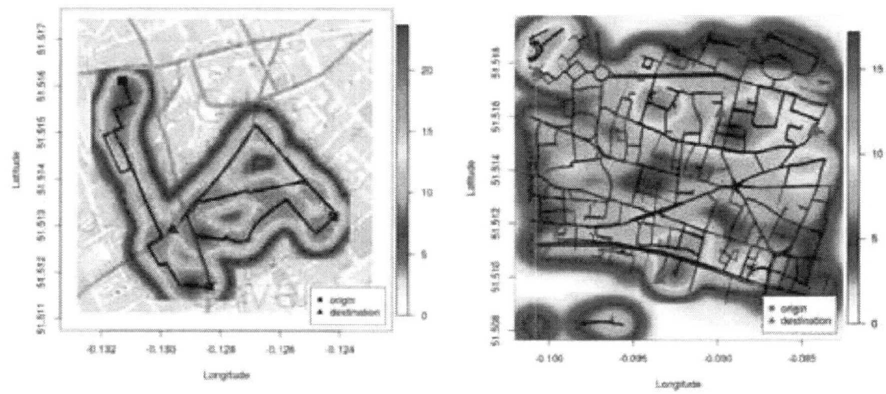

Figure 6. An example of an agent-based model simulating the congestion areas of the London financial district for hypothetical pedestrian flows: left, point-to-point walking behavior from a set of local underground stations (denoted by black squares) to place of work; right, random behavior simulating, e.g., tourist traffic originating from underground stations and converging at local points of interest. Adapted from [49].

distinguish an individual's knowledge of the urban network they are traversing. A person with partial or limited knowledge exhibits a different behavior set compared to a person with full knowledge who can optimize travel based on this and grid perception. Other factors also contribute to decision-making—e.g. personal walking preference (aggressive, cautious, or random), age, pedestrian group size, and so on. As an example, **Figure 7** illustrates age and walking preference-based differences in terms of time taken and distance traveled overall for the financial sector of the City of London grid. In evacuation scenarios, a similar decision state can be used to simulate behavioral aspects under emergency conditions; decision factors can range from group dynamics, placement of safe areas, visual perception under reduced visibility conditions, and "fear." Transition and waiting states aim to simulate the action part of the behavior, namely, execution and re-evaluation as the situation develops. This state transition diagram is illustrated in **Figure 8**.

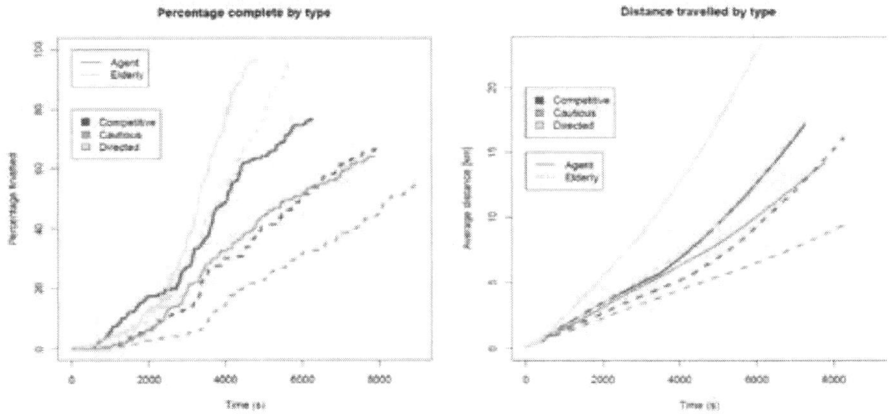

Figure 7. Simulation of age-based agent route performance when traversing an urban network [49], for pedestrians of different behavioral types.

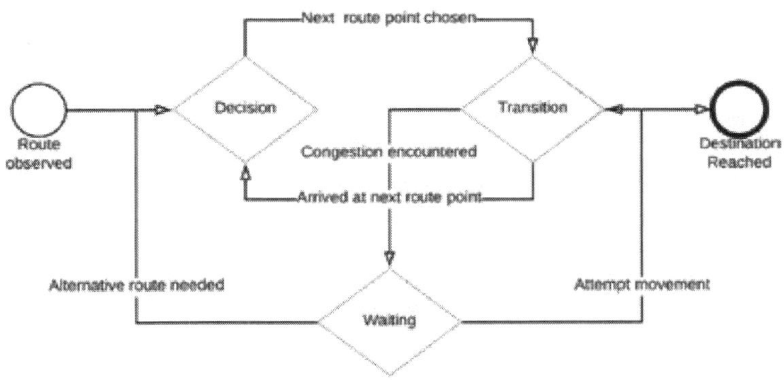

Figure 8. An agent decision-transition-waiting diagram, as implemented by the agent-based model in [49].

This type of model detail facilitates understanding of actual patterns observed in both traffic and pedestrian flows and evacuation scenarios. Similar features can occur for pedestrians to those found in traffic modeling (e.g., lane formation—as agents have to wait for other agents on the same route choice path to move out of the way). Route choice preferences (and those with high throughput) are clearly visible. Moreover, as lanes form, the ABM model allows for re-evaluation of routes based on dynamic parameters like congestion (e.g., during large crowd events, where destination nodes in the urban or evacuation grid display become jammed, usually due to too few approaches, or alternative exits rendered unreachable due to blocking by incomers or slow movers). Congestion avoidance of fellow pedestrians in free move- ment and in crowds is also readily simulated using the agent basis. Pedestrians make optimal choices in the context only of perceived *local* grid congestion (as opposed to global knowledge of congestion points). Finally, ABM allow for clear identification of network *inflection points*, when impact of crowd size on travel or evacuation times becomes exponential rather than linear.

It should be emphasized again that a critical aspect of ABM performance for these problems is the choice of VGI or GIS platforms used to source the grid information. A platform such as OpenStreetMap permits extraction and visualization of relatively accurate street-level details, not only with respect to street geometry and space but also in terms of street metadata such as throughput and physical street characteristics—e.g., length and width of sidewalk, etc. These data are critical to provision of accurate grid simulations and assessment of pressure points and associated risks. Furthermore, the ability to edit the data to permit experimental analysis of the impact of alternative urban layouts and scenarios is important in building a relevant model, with a potential for understanding, anticipating, and responding to a range of pedestrian behavior. Linkage to geographic information systems (GIS), combining spatial and temporal aspects, additionally promises an effective geo-simulation tool facilitating interpretation of the urban environment [95]. Nevertheless, models using *both* separate crowdsourced GIS and ABM are relatively rare [96], and further investigation of social behavior patterns is clearly required.

5. Conclusion

In this chapter, we have discussed factors influencing pedestrian urban mobility, which motivate ongoing research in commute efficiency, together with the wider implications for health and safety. Urban grids with high throughput typically utilize multiple transport modes and require efficient navigation, with non-motorized options increasingly seen as important in terms of reduction of harmful CO_2 emissions and benefits to health. In addition, infrastructure expansion and population growth present increased challenges for city management and emergency responders. Recently, the ability to visualize urban networks with greater accuracy has received considerable impetus from the emergence of new tools, such as VGI platforms, on which detailed simulations can be built. The use of stochastic agent-based models in these simulations has proven particularly useful in terms of evaluating urban layouts and the diverse patterns of pedestrian movement. Moreover, ABM combined with VGI demonstrates considerable potential in modeling a range of real-world situations, ranging from crowd formation and dispersion to evacuation in the event of natural and man-made disasters.

References

1. Vermuyten H, Beliën J, De Boeck L, Reniers G, Tony W. A review of optimisation models for pedestrian evacuation and design problems. Safety Science. 2016;87:167-178. DOI: 10.1016/j. ssci.2016.04.001

2. Ge YE, Long J, Xiao F, Shi Q. Traffic modeling for low-emission transport. Transportation Research Part D: Transport and Environment. 2018;60:1-6. DOI: 10.1016/j.trd.2017.11.001
3. Goetz A. Suburban sprawl or urban centres: Tensions and contradictions of smart growth approaches in Denver, Colorado. Urban Studies. 2013;50(11):2178-2195. DOI: 10.1177/0042098013478238
4. Martellozzo F, Clarke KC. Measuring urban sprawl, coalescence, and dispersal: A case study of Pordenone, Italy. Environment and Planning. B, Planning & Design. 2011;38(6):1085-1104. DOI: 10.1068/b360890
5. Li XH, Liu JL, Gibson V, Zhu YG. Urban sustainability and human health in China, East Asia and south East Asia. Environmental Sustainability. 2012;4(4):436-442. DOI: 10.1016/j. cosust.2012.09.007
6. Chadchan J, Shankar R. An analysis of urban growth trends in the post- economic reforms period in India. International Journal of Sustainable Built Environment. 2012;1(1):36-49. DOI: 10.1016/j.ijsbe.2012.05.001
7. Liu Y, Chen J, Cheng W, et al. Spatiotemporal dynamics of the urban sprawl in a typical urban agglomeration: A case study on southern Jiangsu, China (1983-2007). Frontiers in Earth Science. 2014;8(4):490-504. DOI: 10.1007/ s11707-014-0423-1
8. Sun L, Wei J, Duan DH, Guo YM, Yang DX, Jia C, et al. Impact of land-use and land-cover change on urban air quality in representative cities of China. Journal of Atmospheric and Solar - Terrestrial Physics. 2016;142:43-54. DOI: 10.1016/j. jastp.2016.02.022
9. Wei YD, Ewing R. Urban expansion, sprawl and inequality. Landscape and Urban Planning. 2018;177:259-265. DOI: 10.1016/j.landurbplan.2018.05.021
10. Population dynamics of UK city regions since mid-2011, Office for National Statistics, GB, October 2016. pp. 1-26
11. Tikoudis I, Walid O, Alexandros D, Lara E. Rethinking Urban Sprawl: Moving Towards Sustainable Cities. 2018. DOI: 10.1787/9789264189881-en
12. Mayor of London, Transport for London, TfL, Budget 2018/19, 2018. pp. 1-82
13. Swanwick C, Dunnett N, Woolley H. Nature, role and value of green space in towns and cities: An overview. Built Environment. 2003;29(2):94-106
14. Shiwakoti N, Sarvi M. Understanding pedestrian crowd panic: A review on model organisms approach. Journal of Transport Geography. 2013;26:12-17. DOI: 10.1016/j. jtrangeo.2012.08.002
15. Helbing D, Molnar P, Farkas IJ, Bolay K. Self-organizing pedestrian movement. Environment and Planning. B, Planning & Design. 2001;28:361-383. DOI: 10.1068/ b2697
16. Ma J, Gong WG, Lo SM, Fang M. New insights into turbulent pedestrian movement pattern in crowd- quakes. Journal of Statistical Mechanics: Theory and Experiment. 2013;23:02028. DOI: 10.1088/1742-5468/2013/02/ P02028

17. Duives DC, Wang G, Kim J. Forecasting pedestrian movements using recurrent neural networks: An application of crowd monitoring data. Sensors (Basel). 2019;19(2):382. DOI: 10.3390/s19020382

18. Cantillo V, Arellana J, Rolong M. Modelling pedestrian crossing behaviour in urban roads: A latent variable approach. Transportation Research Part F: Traffic Psychology and Behaviour. 2015;32:56-67. DOI: 10.1016/j.trf.2015.04.008

19. Cochoy F, Hagberg J, Canu R. The forgotten role of pedestrian transportation in urban life: Insights from a visual comparative archaeology. Urban Sudies. 2015;52(12):2267-2286. DOI: 10.1177/0042098014544760

20. Couzin ID, Krause J. Self- organization and collective behavior in vertebrates. Advances in the Study of Behavior. 2003;32:1-75. DOI: 10.1016/ S0065-3454(03)01001-5

21. Laurier E, Brown B, McGregor M. Mediated pedestrian mobility: Walking and the map app. Mobilities. 2016;11(1):117-134. DOI: 10.1080/17450101.2015.1099900

22. Grimm NB, Faeth SH, Golubiewski NE, Redman CL, Wu J, Bai X, et al. Global change and the ecology of cities. Science. 2008;319:756-760. DOI: 10.1126/science.1150195

23. Mestayer P, Abidi A, André M, et al. Urban mobility plan environmental impacts assessment: A methodology including socio-economic consequences-The Eval-PDU project. 10th Urban Environment Symposium, Urban Futures for a Sustainable World. Jun 2010, Göteborg, Sweden. 11p. ffhal- 00614951f. HAL ID: hal-00614951

24. Sustainable Cities, The Aalborg Commitments [Internet]. Available from: http:// www.sustainablecities. eu/fileadmin/repository/Aalborg_ Commitments/Aalborg_ Commitments_ English.pdf [Accessed 11-05-2019]

25. Carriou C, Ratouis O. Is there a model for sustainable urban planning? Series: The Renewed Relevance of Urban-Planning Models? pp. 1-7 [Internet]. Available from: Metropolitiques.eu [Accessed 11-05-2019]

26. World Economic Situation and Prospects. New York, UN: Depart. of Ecomonic and Soc. Affairs; 2013. pp. 1-207

27. Augustijn-Beckers EW, Flacke J, Retsios B. Investigating the effect of different pre-evacuation behavior and exit choice strategies using agent-based modeling. Procedia Engineering. 2009;3:23-35. DOI: 10.1016/j. proeng.2010.07.005

28. Solar Panel In Singapore, COO Publin Domain. [Internet]. Available from:https:// www.publicdomainpictures. net/en/view-image. php?image=32911&picture=solar-panel- in-singapore [Accessed 11-05-2019]

29. Location ETH Zurich, Zentrum. [Internet]. Available from:https:// www.ethz.ch/en/ news-and-events/ media-information/images/standort- eth-zentrum.html [Accessed 11-05-2019]

30. Hoogendoorn S, Daamen W. Self- organization in pedestrian flow. In: Hoogendoorn SP et al., editors. Traffic and Granular Flow '03. Berlin, Heidelberg: Springer; 2005. pp. 373-382. DOI: 10.1007/3-540-28091-X_36

31. Huang L, Xia Y, Wong SC, Shu C-W, Zhang M, Lam WHK. Dynamic continuum model for bi-directional pedestrian flows. Proceedings of the Institution of Civil

Engineers - Engineering and Computational Mechanics. 2009;162(2):67-75. DOI: 10.1680/eacm.2009.162.2.67

32. Duives DC, Daamen W, Hoogendoorn SP. State-of-the-art crowd motion simulation models. Transportation Research Part C: Emerging Technologies. 2013;37: 193-209. DOI: 10.1016/j.trc.2013.02.005

33. Weifeng F, Lizhong Y, Weicheng F. Simulation of bi-direction pedestrian movement using a cellular automata model. Physica A: Statistical Mechanics and its Applications. 2003;321(3-4):633-640. DOI: 10.1016/ S0378-4371(02)01732-6

34. Teknomo K, Gerilla GP. Mesoscopic Multi-Agent Pedestrian Simulation. In: Inweldi PO, editor. Transportation Research Trends. Vol. 1. Nova Science Publishers, Inc; 2008. p. 14. ISBN 978-1-60456-031-2

35. Lee JYS, Lam WHK. Simulating pedestrian movements at signalized crosswalks in Hong Kong. Tansportation Research Part A: Policy and Practice. 2008;42(10):1314-1325. DOI: 10.1016/j. tra.2008.06.009

36. Papadimitriou E, Lassarre S, Yannis G. Introducing human factors in pedestrian crossing behaviour models. Transportation Research Part F: Traffic Psychology and Behaviour. 2016;36: 69-82. DOI: 10.1016/j.trf.2015.11.003

37. Chraibi M, Kemloh U, Schadschneider A, Seyfried A. Force- based models of pedestrian dynamics. Networks & Heterogeneous Media. 2011;6(3):425-442. DOI: DIO:10.3934/ nhm.2011.6.425

38. Moussaïd M, Helbing D, Theraulaz G. How simple rules determine pedestrian behavior and crowd disasters. Proceedings of the National Academy of Sciences. 2011;108(17):6884-6888. DOI: 10.1073/ pnas.1016507108

39. Hoogendoorn SP, Bov PHL. Pedestrian route-choice and activity scheduling theory and models. Transportation Research Part B: Methodological. 2004;38(2):169-190. DOI: 10.1016/S0191-2615(03)00007-9

40. Oxley J, Fildes B, Ihsen E, Charlton J, Days R. Crossing roads safely: An experimental study of age differences in gap selection by pedestrians. Accident Analysis and Prevention. 2005;37:962-971. DOI: 10.1016/j. aap.2005.04.017

41. Mitchell DH, Smith JMG. Topological network design of pedestrian networks. Transportation Research Part B: Methodological. 2001;35(2):107-135. DOI: 10.1016/ S0191-2615(99)00039-9

42. Hughes RL. A continuum theory for the flow of pedestrians. Transportation Research Part B: Methodological. 2002;36(6):507-535. DOI: 10.1016/ S0191-2615(01)00015-7

43. Piccoli B, Tosin A. Archive for Rational Mechanics and Analysis. 2011;199(3):707-738. DOI: 10.1007/ s00205-010-0366-y

44. Degond P, Appert-Rolland C, Pettré J, Theraulaz G. Vision-based macroscopic pedestrian models. Kinetic and Related Models. 2013;6:809-839. DOI: 10.3934/ krm.2013.6.809

45. Papadimitriou E, Yannis G, Golias J. A critical assessment of pedestrian behaviour models. Transportation Research Part F: Traffic Psychology and Behaviour. 2009;12(3):242-255. DOI: 10.1016/j.trf.2008.12.004

46. Blue VJ, Adler JL. Emergent fundamental pedestrian flows from cellular automata microsimulation. Transportation Research Record. 1998;1644(1):29-36. DOI: 10.3141/1644-04

47. Liu R, da Silva JPC, da Maia Seco AJ. A Bi-modal microsimulation tool for the assessment of pedestrian delays and traffic management. In: The Proceedings of the 9th International Association of Travel Behaviour Research Conference, Gold Coast, Australia. 2000

48. Heliövaara S, Korhonen T, Hostikka S, Ehtamo H. Counterflow model for agent-based simulation of crowd dynamics. Building and Environment. 2012;48:89-100. DOI: 10.1016/j. buildenv.2011.08.020

49. Bezbradica M, Ruskin HJ. Modelling impact of morphological urban structure and cognitive behaviour on pedestrian flows. In: Murgante B et al., editors. Computational Science and its Applications – ICCSA 2014. Vol. 8582. Cham: LNCS, Springer; 2014. pp. 268- 283. DOI: 10.1007/978-3-319-09147-1_20

50. Batty M, Multi-Agent Simulation JB. Computational dynamics within GIS. In: Martin D, Artkinson P, editors. Innovations in GIS VII: Geocomputation. London: Taylor & Francis; 2000. pp. 55-71

51. Smart cities. Cities using technological solutions to improve the management and efficiency of the urban environment. European Commission, 2014-2020. Available from: https:// ec.europa.eu/info/eu-regional-and- urban-development/topics/cities-and- urban-development/city-initiatives/ smart-cities_en. Accessed: [11-05-2019]

52. Naghawi H, Wolshon B. Performance of traffic networks during multi-modal evacuations: Simulation- based assessment. Natural Hazards Review. 2012;13(3):196-204. DOI: 10.1061/(ASCE)NH.1527-6996.0000065

53. Lämmel G, Grether D, Kai N. The representation and implementation of time-dependent inundation in large-scale microscopic evacuation simulations. Transportation Research Part C: Emerging Technologies. 2010;18(1):84-98. DOI: 10.1016/j. trc.2009.04.020

54. Burstedde C, Klauck K, Schadschneider A, Zittartz J. Simulation of pedestrian dynamics using a two-dimensional cellular automaton. Physica A: Statistical Mechanics and its Applications. 2001;295(3-4):507-525. DOI: 10.1016/ S0378-4371(01)00141-8

55. Kuligowski ED, Rd P. A Review of Building Evacuation Models. National Institute of Standards and Technology; 2005. p. 156

56. Ioannis K, Nick S, Rahul N, Stephen JG. Implicit crowds: Optimization integrator for robust crowd Simulation. ACM Transactions on Graphics. 2017;36(4):136:1-136:13. DOI: 10.1145/3072959.3073705

57. Tan L, Hu M, Lin H. Agent-based simulation of building evacuation: Combining human behavior with predictable spatial accessibility in a fire emergency. Information Sciences. 2015;295:53-66. DOI: 10.1016/j. ins.2014.09.029

58. Rahman A, Mahmood AK. Agent- based simulation using prometheus methodology in evacuation planning. In: International Symposium on Information Technology, Kuala Lumpur. 2008. pp. 1-8. DOI: 10.1109/ ITSIM.2008.4632061

59. Litman T. Lessons from Katrina and Rita: What major disasters can teach transportation planners. Journal of Transportation Engineering. 2006;132(1):11

60. Kovács G, Spens KM. Humanitarian logistics in disaster relief operations. International Journal of Physical Distribution and Logistics Management. 2007;37(2):99-14. DOI: 10.1108/09600030710734820

61. Future flood losses in major coastal cities. Nature Climate Change. 2013;3:802-806. DOI: 10.1038/ nclimate1979

62. Pillac V, Van Hentenryck P, Even C. A conflict-based path-generation heuristic for evacuation planning. Transportation Research Part B: Methodological. 2016;83:136-150. DOI: 10.1016/j.trb.2015.09.008

63. Cameron GDB, Duncan GID. The Journal of Supercomputing. 1996;10:25. DOI: 10.1007/BF00128098

64. Halati A, Lieu H, Walker S. CORSIM - corridor traffic simulation model. In: Traffic Congestion and Traffic Safety in the 21st Century: Challenges, Innovations, and Opportunities. Chicago, Illinois; 1997. pp. 570-576

65. Fellendorf M, Vortisch P. Microscopic traffic flow simulator VISSIM. In: Barceló J, editor. Fundamentals of Traffic Simulation. International Series in Operations Research & Management Science. Vol. 45. New York, NY: Springer; 2010. DOI: 10.1007/978-1-4419-6142-6_2

66. Tu H, Tamminga G, Drolenga H, de Wit J, van der Berg W. Evacuation plan of the city of Almere: Simulating the impact of driving behavior on evacuation clearance time. Procedia Engineering. 2010;3:67-75. DOI: 10.1016/j.proeng.2010.07.008

67. Ronchi E, Gwynne SMV, Rein G, Wadhwani R, Intini P, Bergstedt A. e-Sanctuary: Open Multi-Physics Framework for Modelling Wildfire Urban Evacuation. Project Report. National Fire Protection Association; 2017

68. City of Chicago, Central Business District Evacuation Plan. Published by Chicago Police Department on 16-07-2005

69. OpenStreetMap, [Internet]. Available from:https://www. openstreetmap.org. [Accessed 11-05-2019]

70. O'Flaherty CA, Parkinson MH. Movement on a city Centre footway. Traffic Engineering and Control. 1972;13:434-438

71. Kirchner A, Klüpfel H, Nishinari K, Schadschneider A, Schreckenberg M. Discretization effects and the influence of walking speed in cellular automata models for pedestrian dynamics. Journal of Statistical Mechanics: Theory and Experiment. 2004;10:10011. DOI: 10.1088%2F1742- 5468%2F2004%2F10%2Fp10011

72. Pan X, Han CS, Dauber K, Law KH. A multi-agent based framework for the simulation of human and social behaviors during emergency evacuations. AI & Society. 2007;22(2):113-132. DOI: 10.1061/40794(179)92

73. Lu Q , George B, Shekhar S. Capacity constrained routing algorithms for evacuation planning: A summary of results. In: Bauzer Medeiros C, Egenhofer MJ, Bertino E, editors. Advances in Spatial and Temporal Databases SSTD 2005. Lecture Notes in Computer Science. Vol. 3633. Berlin, Heidelberg: Springer; 2005. pp. 291-307. DOI: 10.1007/11535331_1F

74. Moussaïd M, Perozo N, Garnier S, Helbing D, Theraulaz G. The walking behaviour of pedestrian social groups and its impact on crowd dynamics. PLoS One. 2010;5(4):1-7. DOI: 10.1371/ journal.pone.0010047

75. Duckham M, Kulik L. "Simplest" paths: Automated route selection for navigation. In: Kuhn W, Worboys MF, Timpf S, editors. Spatial Information Theory, Foundations of Geographic Information Science COSIT 2003. Lecture Notes in Computer Sciences. Vol. 2825. Berlin, Heidelberg: Springer; 2003. pp. 169-185. DOI: 10.1007/978-3-540-39923-0_12

76. Hillier B, Iida S. Network and psychological effects in urban movement. In: Cohn AG, Mark DM, editors. Spatial Information Theory COSIT 2005. Lecture Notes in Computer Sciences. Vol. 3693. Berlin, Heidelberg: Springer; 2005. pp. 475-490. DOI: 10.1007/11556114_30

77. Fujii H, Uchida H, Yoshimura S. Agent-based simulation framework for mixed traffic of cars, pedestrians and trams. Transportation Research Part C: Emerging Technologies. 2017;85:234-248. DOI: 10.1016/j.trc.2017.09.018

78. Crooks A, Castle C, Batty M. Key challenges in agent-based modelling for geospatial simulation. Computers, Environment and Urban Systems. 2008;32(6):417-430. DOI: 10.1016/j. compenvurbsys.2008.09.004

79. Antonini G, Bierlaire M, Weber M. Discrete choice models of pedestrian walking behavior. Transportation Research Part B: Methodological. 2006;40(8):667-687. DOI: 10.1016/j. trb.2005.09.006

80. Helbing D, Johansson A, Al-Abideen ZH. Crowd turbulence: The physics of crowd disasters. In: The Fifth International Conference on Nonlinear Mechanics (ICNM-V). 2007. pp. 967-969 arXiv:0708.3339

81. Helbing D, Mukerji P. Crowd disasters as systemic failures: Analysis of the love parade disaster. EPJ Data Science. 2012;1. DOI: 10.2139/ ssrn.2191620

82. Aghabayk K, Ejtemai O, Sarvi M, Sobhani A. Understanding pedestrian crowd merging. Transportation Research Procedia. 2014;2:768-773. DOI: 10.1016/j.trpro.2014.09.086

83. Weiss T, Litteneker A, Jiang C, Terzopoulos D. Position-based Multi- Agent dynamics for real-time crowd Simulation - MiG 2017. In: Proceedings of MiG '17, Barcelona, Spain, November. Vol. 9. 2017. pp. 8-10. DOI: 10.1145/3136457.3136462

84. D'Orazio M, Spalazzi L, Quagliarini E, Bernardini G. Agent-based model for earthquake pedestrians' evacuation in urban outdoor scenarios: Behavioural patterns definition and evacuation paths choice. Safety Science. 2014;62:450-465. DOI: 10.1016/j.ssci.2013.09.014

85. Gilbert N, Troitzsch KG. Simulation of the Social Scientist. 1st Ed. Buckingham. Vol. 308. Philadelphia: Open University Press; 1999

86. Fiedrich F, Burghardt P. Agent- based Systems for Disaster Management. Communications of the ACM. 2007;50(3):41-42. DOI: 10.1145/1226736.1226763

87. Castle CJE, Crooks AT. Principles and Concepts of Agent-Based Modelling for Developing Geospatial Simulations. London: CASA Working Papers, Centre for Advanced Spatial Analysis, UCL; 2006. p. 110

88. Schelling TC. Dynamic models of segregation. Journal of Mathematical Sociology. 1971;1:143-186. DOI: 10.1080/0022250X.1971.9989794

89. Cetin N, Nagel K, Raney B, Voellmy A. Large-scale multi- agent transportation simulations. Computer Physics Communications. 2002;147:559-564. DOI: 10.1016/S0010-4655(02)00353-3

90. Omer I, Kaplan N. Using space syntax and agent-based approaches for modeling pedestrian volume at the urban scale. Computers, Environment and Urban Systems. 2017;64:57-67. DOI: 10.1016/j. compenvurbsys.2017.01.007

91. Ueno J, Nakazawa A, Kishimoto T. An analysis of pedestrian movement in multilevel complex by space syntax theory: In the case of Shibuya Station. Journal of Civil Engineering and Architecture. 2009;118:1-12. DOI: 10.17265/1934-7359/2012.07.002

92. Raford N. Social and technical challenges to the adoption of space syntax methodologies as a planning support system (PSS) in American Urban Design. The Journal of Space Syntax. 2010;1(1):230-245

93. Henein CM, White T. Agent- based modelling of forces in crowds. In: Davidsson P, Logan B, Takadama K, editors. Multi-Agent and Multi- Agent-Based Simulation MABS 2004. Lecture Notes in Computer Sciences. Vol. 3415. Berlin, Heidelberg: Springer; 2005. pp. 173-184. DOI: 10.1007/978-3-540-32243-6_14

94. Longley PA, Batty M, editors. Advanced Spatial Analysis: The CASA Book of GIS. Redlands (USA): ESRI Press; 2003. p. 464

95. Crooks AT, Castle CJE. The integration of Agent-based modelling and geographical information for geospatial Simulation. In: Heppenstall AJ, Crooks AT, See LM, Batty M, editors. Agent-Based Models of Geographical Systems. Dordrecht (Netherlands): Springer; 2012. pp. 219-251. DOI: 10.1007/978-90-481-8927-4_12

96. Andrew TC, Sarah W. GIS and agent-based models for humanitarian assistance. Computers, Environment and Urban Systems. 2013;41:100-111. DOI: 10.1016/j. compenvurbsys.2013.05.003

Chapter-10

Dynamic Street Parking Space Using Memetic Algorithm for Optimization

1. Introduction

As a result of increasing industrialization and urbanization, cities around the world have experienced growing populations and as a result increasing number of vehicles. This trend is growing exponentially. Many predictors of urban population have indicated that by 2050, it is expected that 70% of the world population will be living in cities [1]. This implies that the number of vehicles will also see exponential growth as people work and study in cities. However, while the parking spaces are expanded, it does not meet the demands of growing populations. The shortage of parking spaces exerts more pressure on congested roads, which affects the travel time of motorists. This congestion of traffic leads to increase in costs in terms of fuel consumption, as well as an increase in accidents. The traffic authorities cannot cope with this burgeoning problem since it requires re-designing the parking infrastructure to make it more futuristic, so as to cater for rapid development and exponential growth in the number of motor vehicles [2]. In congested roads, the search for parking space consumes a lot of time and this negatively impacts on the economy, as people spend chunks of time waiting for limited parking spaces. This time would be well spent doing beneficial work. Moreover, the economy is also affected with respect to the environment as more carbon emissions are produced in the course of searching for parking [3]. This adds more strain on the economy as it has to deal with the effects of carbon pollution. The barrels of oil that are consumed everyday as a result of parking search are estimated at about one million rand [4]. Motorists also become fatigued from this tiring exercise of waiting and looking for parking, which affects productivity and overall health of people. Although many approaches have been proposed to tackle this problem [5], more intelligent mechanisms are needed to solve this critical problem, especially given its complexity and scalability. In many cities, parking attendants are ubiquitous as they guide motorists to empty parking bays. However, these manual controls are not efficient and do not optimize the available parking space. They are also expensive since these attendants expect a monetary reward for their services, and in many cases, they demand high charges for their unsolicited and informal services [6]. This is in addition to the formal fee that motorists pay for parking. If the informal attendants are not paid, they might damage the car or slash the tires, which creates a security risk to the motorists. As a result of this car parking problem, a safety hazard is created as carjackers and other robbers disguise themselves as parking attendants and end up inflicting harm on innocent motorists that are desperate for parking space. Due to the far-reaching impact of this problem, policymakers in some cities have attempted to introduce measures of dealing with the issue, by creating more off-street parking spaces and increasing the charges for on-street parking [7]. However, these measures have not created the desired effect as determined from many surveys; most motorists prefer on-street parking [8]. This is due to convenience and proximity of parking space to the intended destination. A long walk to the destination might impel the motorist to jostle and wait for on-street parking space. Therefore, the solution is not to walk away from reality on the off chance that creating more off-street parking would ameliorate the problem, but to face the problem and find smarter ways of optimiz- ing the on-street parking. It is from that premise that the idea of a hybrid optimi- zation method is proposed in this chapter. The method applies memetic algorithm, which is a blend of two algorithms. These are genetic algorithm that is used to optimize the global search based on the natural evolutionary process of generating new offspring using a set of parameters [9] and tabu search (TS) that is a metaheuristic that is used to diversify the search space by conducting neighborhood search of the individual solutions with a view to improve their fitness [10]. Unlike other heuristics, TS is memory based, such that each found solution in the course of the iteration is stored in a tabu list, so as to keep track of solutions and avoid duplicates. Its adaptive memory makes it suitable for highly dynamic problems [11], such as the street parking problem (SPP). In order to generate an optimal solution for the SPP, the synergy of these algorithms is harnessed to attain intensification and diversification search processes and enable the search reach optimum without premature convergence. The remainder of this chapter is structured as follows: related work is presented in Section 2, problem definition is talked about in Section 3, methodology is discussed in Section 4, results and discussion make Section 5, and the article concludes in Section 6.

2. Related work

A number of evolutionary algorithms have been advanced in the literature in an effort to solve the street parking problem (SPP). The review of these approaches provides an insight into what has been done and the potential impact and utility of the proposed approach in the current body of related knowledge.

2.1. Single metaheuristics

Thomas and Kovoor [12] applied genetic algorithm to create an autonomous system for parking vehicles. The approach used the roulette wheel selection method for selection of mating individuals and then used genetic operators, such as cross- over and mutation to optimize fitness and produce offspring. The same principle was used by Aydin et al. [13] to develop a parking system for a smart city. Based on Internet of Things, data is transmitted across the Internet through devices. The genetic algorithm generates a vacant parking slot, which can be navigated for reservation by the motorist. This process can be done on a mobile device, which is similar to the intelligent parking system that operates on android platform [14]. In order to automate the parking guide process, a mobile robot was designed to optimize parking space using fuzzy logic [15]. The intelligent mobile robot automatically guides the motorist toward free parking space. Based on genetic algo- rithm, the robot is able to detect free parking space and map it to the vehicle. It also checks the dimensions of the parking bay to ensure an appropriate vehicle is assigned the bay. The deep learning of the robot system creates patterns that can be used for prediction; for example, Camero et al. [16] applied neural networks to predict the occupancy of the parking, so as to output the possible free parking spaces available through an optimization process. Han et al. [17] optimized the selection of the parking lot and route by use of Dijkstra's algorithm. The outcome of the optimization process in terms of path and parking lot is sent to the cell phone of the driver. The travel and walking distances are considered before the automatic assignment of parking.

2.2. Hybrid metaheuristics

Dong et al. [18] hybridized the chaotic dynamics with the particle swarm opti- mization (PSO) to optimize the parking space. In order to avoid premature conver- gence, there is a constant update of the rules with a view to increase diversity and expand the search. This technique was also applied by Abidi et al. [19] whereby parking space is assigned to groups of drivers using a hybrid genetic algorithm.

However, with the changing optima in the current problems related to car parking, a dynamic application of the search operators would make the methods more adaptive, so as to perform well despite the complexity and nonstationary nature of the car parking problem. Researchers have resorted to combining several algorithms to match the inherent difficulty of the problem. For example, Kantenti et al. [20] combined sensor technology with the Internet of Things. The barricade to the parking area is sensor enabled, so as to capture the details of the car including number plate and keep track of the vehicles, as well as both allocated and unallocated parking spaces. The nearest parking space is allocated to the vehicle.

Sensors that are triggered when a vehicle enters or exits the parking lot include Arduino, ultrasonic sensor, and optical character recognition, among others. In order for the vehicle to be allocated the nearest parking space, a swarm intelligence mechanism was applied by Moorthy and Pabitha [21], whereby the route generator and allocator are infused into the PSO algorithm, so as to have source and destina- tion routes generated and allocated to vehicles searching for parking space. Motor- ists request for parking through a request handler function that then sends the information to the route generator. The storage of this information is done by the route repository, and this information can be useful for prediction; for example, Rad et al. [22] use the data collected on vehicles for traffic flow prediction. The approach combines genetic algorithm with neural fuzzy technique for reservation of parking space. The most optimal parking space is generated using a multi-objective function that simultaneously applies genetic and fuzzy operators to the search optimization process with a view to minimize the waiting time and cost that motorists can spend on the parking space. The advent of electric cars has added another layer of difficulty to the parking space problem not only due to increase in cars but also waiting time concerns, as it affects the depletion of the battery power. Amini et al. [23] proposed a power distribution system to support parking lots for electric cars. Genetic algorithm is hybridized with particle swarm optimization algorithm to minimize loss of distribution network and optimize the allocation of parking space. The searching time for parking space can also be minimized through application of a bi-level genetic algorithm that was presented by Shen et al. [24].

The approach finds inter alia, free parking space, fees, and location, which are defined at different levels of the problem. For example, the location is produced by the upper level, while waiting and travel costs are generated by the lower level. The method aims to minimize the costs associated with obtaining the parking space.

3.Problem definition

The problem in this section is defined as an on-street parking optimization problem based on linear programming. This implies that subject to a set of con- straints, the best solution is obtained from a set of feasible solutions [25] with regard to parking space. The objective function is used to calibrate the outcome of the optimization process according to the desired objective or target as per the problem features, such as maximizing or minimizing based on the decision variables and coefficients (Table 1).

Minimize:

$$\sum_{i}^{n}\sum_{j}^{p} C_{ij} D_{ij} T_{ij} W_{ij} \tag{1}$$

Subject to:

$$\sum_{i}^{p} w_{ij} \leq S_j \quad \forall j = 1, 2, \ldots p \quad \forall i = 1, 2, \ldots n \tag{2}$$

$$\sum_{i}^{n} w_{ij} > 0 \quad \forall j = 1, 2, \ldots p \quad \forall i = 1, 2, \ldots n \tag{3}$$

$$W_{ij} = \begin{cases} 1 \text{ if parking space } j \text{ is assigned to car } i \\ 0 \text{ otherwise} \end{cases} \tag{4}$$

Table 1. Meaning of notations used in this section.

Notation	Denotation
i	Car entity: i = 1, 2, ... n is cars available for parking.
j	Space entity: *j* = 1, 2, ... *p* is parking space available.
Cij	Cost of assigning car *i* to parking space *j*
Dij	Distance between car *i* and parking space *j*
Tij	Time it takes to assign car *i* to parking space *j*
Sj	Parking space capacity

The objective of the function defined in Eq. (1) is to minimize the following:

i. The cost of assigning a car to an empty parking space
ii. Distance between intended destination and parking
iii. Time it takes to assign car to an empty parking space, as well as to find empty space and map it onto the waiting car
iv. Vehicle waiting time

In order to find the optimal solution, the objective function is defined with inequality constraints. The constraint in Eq. (2) indicates that the total number of assigned cars must not exceed the size or capacity of the available parking space. The vehicle cannot be assigned parking unless there is an empty parking space, as defined in Eq. (3). In order to make it easy for the proposed hybrid algorithm to solve this problem, a penalty function is applied [26]. The function in Eq. (1) is converted into unconstrained optimization problem.

Minimize:

$$\sum_{i}^{n}\sum_{j}^{p} C_{ij}D_{ij}T_{ij}W_{ij} + \alpha_g A_i + \beta_g A_2 \tag{5}$$

A constant penalty is applied to individuals or solutions in the population sample that have violated the constraints. The penalty factors are defined by α_g, $\beta_g \gamma_g$, whereas A_1, A_2, A_3 can be described as follows:

$$A_1 = \sum_{g=1}^{n} \alpha_g * \max\left(0, \left|\sum_{i=1}^{n} w_{ij} - s_{ij}\right|\right)^2 \tag{6}$$

$$A_2 = \sum_{g=1}^{p} \beta_g * \max\left(0, \left(p_J - \sum_{J=1}^{P} w_{ij}\right)\right)^2 \tag{7}$$

where p_J is the parking space and for i 1, 2, …, n;j 1, 2, …, p:

4. Methodology

4.1. Memetic algorithms (MAs)

Unlike genetic algorithms (GAs), MAs combine two evolutionary processes to solve problems. These include the natural process where genes are transmitted from one entity to the other through mating

```
Generate a set of parking space vectors P ← {λ1,λ2, … , λN,}
Define the capacity of each parking space
Get the neighborhood, H(j) of each parking space vector λi,i = 1,2, … , N
PO ← Initialize the population()
Evaluate fitness
while stop criterion is not reached do
  fori = 1,2, … , N  do
   if vehicle,v > 0 then
 P(j) ← H(j)
      Apply crossover
     Apply mutation
     Apply recombination
v ← P(j)
   else
P(j) ← {1,2, … , N}
   end if
 Adjust penalty parameter,δ
  Update current population
 end for
   P ← LocalSearch(P)
end while
```

Figure 1. Memetic algorithms.

and the cultural process where memes are transmitted through meeting and interactions with one another [27] (Figure 1).

These steps are followed in the implementation of the proposed algorithm. The algorithm represents a hybridized structure of evolutionary method and the local search. Memetic operators such as recombination and mutation are applied, so as to re-create new solutions through interactions between each other and the environ- ment with a view to generate a better offspring. The local search expands the search space, which increases population diversity. The presence of multiple memes provides an opportunity for fitness comparisons, so that the best outcome is returned by the algorithm [28]. Whereas most approaches use tournament and roulette selection methods for parent individuals in the population, a random selection mechanism is used in this work [29]. The random selection of parents is conducted with a view to give each individual an equal chance for selection. The memetic operators are then applied to produce offspring. MAs are considered one of the best evolutionary methods that can solve today's increasingly complex and dynamic problems and because of this, they have been extensively applied to solve real-life problems [30, 31].

4.2. Tabu search (TS)

In this work, TS metaheuristic is used to perform the exploitation procedure or search the local neighborhood for better solutions. TS comprises in its structure the ability to store solutions in the course of iterations [11]. Based on its memory function, TS prevents duplication of solutions by creating a sequential tabu list of solutions. The solution is added on condition that it is better than the existing previous solution [32]. The feasibility of the solutions is maintained by penalizing the infeasible solutions and concentrating the search in promising areas of the search space. The TS helps to control the search, given that it is easier to know which areas contain feasible solutions or otherwise based on the tabu list; the search can then be directed to areas of high feasibility value with a view to obtain the best solutions. The optimal parking space for the vehicle is thus obtained through search intensification and diversification. An adaptive memory configuration for TS is considered in this work. This allows automatic adjustment of the tabu list as the search progresses. The search moves that were already made or the solutions that were visited are forbidden, so as to ensure that all areas are visited during the search [33]. Swap moves are employed to focus search attention to areas that do not feature in the tabu search list. A swap move is conducted if it will yield the highest move gain [34]. The variables cannot change their status once they have been included in the tabu list, but if the

```
Start with a solution, S_0,
Search the neighborhood of S_0,
while stop criteria is not reached
  consider the constraints, equations 2-3
  Get the best solution, p(S_0^*)
 if p(S_0^*) > p(S_0,)
   Record the current move in the tabu list, T
   Replace p(S_0,) with p(S_0^*)
 else
   Keep solution unchanged
 end if
   Output optimal solution
end while
```

Figure 2. Search technique.

search leads to a potentially better solution, then a swap move is performed. These measures enable the local neighborhood search to converge to solutions of good quality [35]. The local search technique implementa- tion in this chapter follows the steps in Figure 2.

5. Results and discussion

The efficiency of the proposed method, adaptive memetic algorithm (AMA), was evaluated against the existing optimization techniques in the literature with a view to investigate the performance of each method over the same set of data and defined with similar settings. The optimization techniques implemented in this work include simulated annealing (SA) [36], genetic algorithm (GA) [37], ant colony optimization (ACO) [38], and particle swarm optimization (PSO) [39].

MATLAB environment was used for the experiments in this section. The case study was taken from the city of Durban in South Africa which has an on-street parking problem (OPP) due to limited parking spaces. Anton Lembede street and Prince street are the two busy streets in Durban city that were selected for this study, and the global positioning system (GPS) data are drawn from the geographical infor- mation system of Google Maps, including the parking space coordinates. Thirty individuals are considered for the population sample size (Figures 3 and 4).

Crossover and mutation probabilities are set at 25 and 75%, respectively. The performance is measured based on the average waiting time, utilization of the parking space, and allocation of appropriate parking space to the car.

The proposed AMA demonstrates the ability to explore and exploit the search better in search of higher fitness individuals during the search process, for most of the instances as shown in Figure 5.

The results indicate that parking space wastage is minimized when memetic algorithm is employed in optimization of parking space. Also, the proposed method, AMA, produces a faster convergence for most iterations as shown in Figure 6. This is due to the local search technique that adaptively conducts a neighborhood search for better solutions. Ultimately, the AMA ensures that an appropriate parking space is allocated to the car.

Figure 3. Satellite view of the Anton Lembede street in Durban city.

Figure 4. Satellite view of the Prince Street in Durban city.

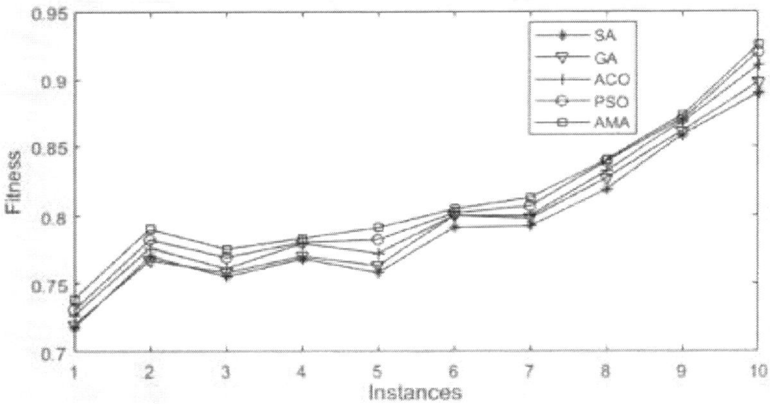

Figure 5. Fitness comparisons between the proposed AMA and existing algorithms.

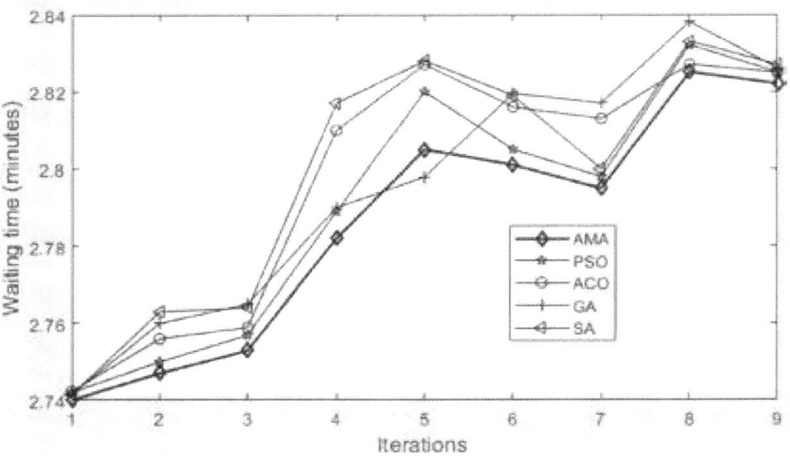

Figure 6. Convergence comparisons between different techniques.

6. Conclusion

In this chapter, an adaptive memetic algorithm was employed to solve the on- street parking problem (OPP). The proposed method combined genetic algorithm (GA) and tabu search (TS) to generate the optimal solution. The method was applied to a real life problem in Durban city of South Africa whose limited parking space is under the control of eThekwini municipality. TS was used to perform neighborhood search to improve the solutions with a view to prevent premature convergence of the search process. The proposed algorithm optimized the parking space search and allocation by reducing the waiting time, cost of allocating vehicle to parking space, and enhancing parking space utilization through appropriate allocations. Given the difficulty of finding and allocating an on-street parking space, the proposed algorithm implemented the memetic operators adaptively, so as to deal with nonstationary optima. The proposed method was compared with existing algorithms, such as SA, GA, ACO, and PSO in solving the same problem, OPP, and as shown in the experiments, the proposed algorithm outperformed all algorithms on several iterations.

References

1. Gross M. The urbanization of our species. Current Biology. 2016;26: 1205-1208. DOI: 10.1016/j.cub.2016. 11.039

2. Yang S, Huang L. Research on planning and management of urban parking lot—taking Hangzhou as an example. Current Urban Studies. 2017;5: 379-386. DOI: 10.4236/cus.2017.54021

3. Bharadwaj S, Ballare S, Rohit R, Chandel MK. Impact of congestion on greenhouse gas emissions for road transport in Mumbai metropolitan region. Transportation Research Procedia. 2017;25:3528-3551. DOI: 10.1016/j.trpro.2017.05.282

4. Gupta A, Kulkarni S, Jathar SV, Jain N. Smart car parking management system using IoT. American Journal of Science, Engineering and Technology. 2017;2:112-119. DOI: 10.11648/j. ajset.20170204.13

5. Abidi S, Krichen S, Alba E, Bravo JM. A hybrid heuristic for solving a parking slot assignment problem for groups of drivers. International Journal of Intelligent Transportation Systems Research. 2017;15:85-97. DOI: 10.1007/ s13177-016-0123-1

6. Foster J, Chasomeris M. Examining car guarding as a livelihood in the informal sector. Local Economy: The Journal of the Local Economy Policy Unit. 2017;32:525-538. DOI: 10.1177/ 0269094217727990

7. Brooke S, Quddus M, Ison S. On- street parking search: A UK local authority perspective. Journal of Transport and Land Use. 2017;10:13-26. DOI: 10.5198/jtlu.2015.600

8. Brooke SL. Factors Influencing Urban on-Street Parking Search Time Using a Multilevel Modelling Approach. Loughborough, England: Loughborough University; 2015

9. Kumari J, Dubey AK. A review paper on genetic algorithm. International Journal of Advance Research in Computer Science and Management Studies. 2016;4:122-125

10. McKendall A, Li C. A tabu search heuristic for a generalized quadratic assignment problem. Journal of Industrial and Production Engineering. 2016;34:221-231. DOI: 10.1080/ 21681015.2016.1253620

11. Xia Y, Fu Z, Pan L, Duan F. Tabu search algorithm for the distance- constrained vehicle routing problem with split deliveries by order. PLoS One. 2018;13:1-19. DOI: 10.1371/ journal. pone.0195457

12. Thomas D, Kovoor BC. A genetic algorithm approach to autonomous smart vehicle parking system. Procedia Computer Science. 2018;125:68-76. DOI: 10.1016/j. procs.2017.12.011

13. Aydin L, Karakose M, Karakose E. A navigation and reservation based smart parking platform using genetic optimization for smart cities. In: 5th International Istanbul Smart Grid an Cities Congress and Fair (ICSG '17); 19-21 April 2017; Istanbul, Turkey: IEEE; 2017. pp. 120-124

14. Anitha J, Thoyajakshi Y, Ramya A, Sravani V, Kumar P. Intelligent parking system using android application. International Journal of Pure and Applied Mathematics. 2017;114:165-174

15. Aye YY, Watanabe K, Maeyama S. An automatic parking system using an optimized image-based fuzzy controller by genetic algorithms. Artificial Life and Robotics. 2017;22:139-144. DOI: 10.1007/s10015-016-0326-1

16. Camero A, Toutouh J, Stolfi DH, Alba E. Evolutionary deep learning for car park occupancy prediction in smart cities. In: Proceedings of the 12th International Conference on Learning and Intelligent Optimization (LION '18); 10-15 June 2018; Kalamata, Greece: LION; 2019. pp. 1-16

17. Han Y, Shan J, Wang M. Optimization design and evaluation of parking route based on automatic assignment mechanism of parking lot. Advances in Mechanical Engineering. 2017;9:1-9. DOI: 10.1177/ 1687814017712416

18. Dong N, Fang X, Wu A-G. A novel chaotic particle swarm optimization algorithm for parking space guidance. Mathematical Problems in Engineering. 2016;2016:1-14. DOI: 10.1155/2016/ 5126808

19. Abidi S, Krichen S, Alba E, Molina JM. A new heuristic for solving the parking assignment problem. Procedia Computer Science. 2015;60:312-321. DOI: 10.1016/j. procs.2015.08.132

20. Kantenti D, Srikar DV, Ramesh TK. Intelligent smart parking problem. In: Proceedings of the IEEE International Conference on Smart Technologies for Smart Nation (SmartTechCon '17); 17-19 August 2017; Bangalore, India: IEEE; 2017. pp. 1-5

21. Moorthy R, Pabitha P. Novel car parking system using swarm intelligence. International Journal of Pure and Applied Mathematics. 2017; 117:289-298

22. Rad F, Pazhokhzadeh H, Parvin H. A smart hybrid system for parking space reservation in VANET. Journal of Advances in Computer Engineering and Technology. 2017;3:12-18

23. Amini HM, Moghaddam MP, Karabasoglu O. Simultaneous allocation of electric vehicles' parking lots and distributed renewable resources in smart power distribution networks. Sustainable Cities and Society. 2017;28: 332-342. DOI: 10.1016/j.scs.2016.10.006

24. Shen X, Chen F, Su B, Chen Q, Yao J. Optimization of park-and-ride system: A case study of Shunyi in Beijing. Advances in Mechanical Engineering. 2017;9:1-8. DOI: 10.1177/ 1687814017714987

25. Rama S, Srividya S, Bellatti D. A linear programming approach for optimal scheduling of workers in a transport corporation. International Journal of Engineering Trends and Technology. 2017;45:482-487

26. Kulkarni O, Kulkarni N, Kulkarni AJ, Kakandikar G. Constrained cohort intelligence using static and dynamic penalty function approach for mechanical components design. International Journal of Parallel, Emergent and Distributed Systems. 2018;33:570-588. DOI: 10.1080/ 17445760.2016.1242728

27. Panteleev AV, Pismennaya VA. Application of a memetic algorithm for the optimal control of bunches of trajectories of nonlinear deterministic systems with incomplete feedback. Journal of Computer and Systems Sciences International. 2018;57:25-36

28. Wang C, Ji Z, Wang Y. A novel memetic algorithm based on decomposition for multiobjective flexible job shop scheduling problem. Mathematical Problems in Engineering. 2017;2017:1-20. DOI: 10.1155/2017/ 2857564

29. Wang C, Tian N, Ji Z, Wang Y. Multi-objective fuzzy flexible job shop scheduling using memetic algorithm. Journal of Statistical Computation and Simulation. 2017;87:2828-2846. DOI: 10.1080/00949655.2017.1344846

30. Pelaez JI, Ruiz JA, Veintimilla J, Vaccaro G, Witt P. Memetic computing applied to the design of composite materials and structures. Mathematical Problems in Engineering. 2017;2017: 1-16. DOI: 10.1155/2017/4723863

31. Wang P, Liu R, Jiang Z. A memetic algorithm for optimization of combination chemotherapy with dose adjustment. In: Proceedings of the 13th IEEE Conference on Automation Science and Engineering (CASE '17); 20-23 August 2017; Xi'an, China: IEEE; 2017. pp. 207-212

32. Niu Y, Yang Z, Chen P, Xiao J. A hybrid tabu search algorithm for a real- world open vehicle routing problem involving fuel consumption constraints. Complexity. 2018;2018:1-12. DOI: 10.1155/2018/5754908

33. Kurt NE, Sahin BH, Kiirsad D. An adaptive tabu search algorithm for market clearing problem in Turkish day-ahead market. In: Proceedings of the 15th International Conference on the European Energy Market (EEM '18); 27-29 June 2018; Lodz, Poland: EEM; 2018. pp. 1-6

34. Wang Y, Wu Q, Punnen AP, Glover F. Adaptive tabu search with strategic oscillation for the bipartite Boolean quadratic programming problem with partitioned variables. Information Sciences—Informatics and Computer Science, Intelligent Systems, Applications: An International Journal. 2018;450:284-300. DOI: 10.1016/j.ins.2018.03.045

35. Roesener AG, Barnes WJ. An advanced tabu search approach to the dynamic airlift loading problem. Logistics Research. 2016;9:1-18. DOI: 10.1007/s12159-016-0139-6

36. Shirdel GH, Abdolhosseinzadeh M. A simulated annealing heuristic for the traveling salesman problem. Journal of Information and Optimization Sciences. 2018;39:1283-1296. DOI: 10.1080/ 02522667.2017.1367494

Chapter-11

Cost-Benefit Evaluation Tools on the Impacts of Transport Infrastructure Projects on Urban Form and Development

1. Introduction

Following the growing world population and associated need for increased transportation, there has been growing interest in some aspects of transport policy such as sustainability in infrastructure provision. The increasing rates of fossil fuel dependency, global warming, poverty and social inclusion are highly relevant for the transport sector [1]. Accordingly, there appears to be a common policy consensus on the need to achieve socio-economic development and environmental protection. Urban development policies internationally now increasingly depend on sustainable transport systems, which include increasing shares of walking, cycling and public transport such as metro and light rail systems [2].

Cost-benefit analysis (CBA) is a tool conventionally used to evaluate potential costs and benefits of a given public investment such as major rail or road invest- ments. The CBA methodology is based on early welfare economic theory [3] where the idea is to achieve efficient allocation of resources and maximise public benefits for general social welfare. In the CBA approach, the effects of public investments are valued in monetary terms and expressed as costs and benefits to represent an overall aggregated value of individual well-being. This approach provides a basis for the evaluation of impacts of various transport projects and policy changes, particularly the rail infrastructure investments in a wide variety of the appraisal work [4]. From this literature, the CBA methodology applied to transport project appraisals consists of a quantification of changes in user costs, infrastructure costs and selected external costs. These are expressed as monetary values, market prices and shadow prices which are then used for the valuation of benefits. The overall result gives guidance on which investment and which alternative are most feasible according to the selected economic criteria.

Based on its wide coverage of impacts and indicators utilised in the appraisal framework, the CBA methodology has increasingly been used as a key tool for a com- prehensive assessment of the impacts of major transport investments and associated policy changes not only in the European Union (EU) but also overseas countries such as US, New Zealand and Australia [5]. Regarding the EU countries, the European Commission's (EC) existing European Research on Transport (EURET) programme has constructed a common basis for research on European transportation-related issues prior to the development of a 'common transport policy'—the Trans-European Network (TEN) programme 2007–2013. This programme aims for more balanced spatial development across the EU by assigning greater importance to regional interconnection, interoperation and access to national and international networks.

Within the context of the aims of TEN, EURET proposes a wide appraisal spectrum for the evaluation of transportation investments across EU countries.

For EU member states, CBA is required for receiving funding from Cohesion Fund or Structural Funds [6]. In a previous EURET Report, it is noted that a majority of European countries rely heavily on CBA in their national appraisal strategies (EURET Concerted Action 1.1., see [7]). The report provided a basis for establishing common appraisal guidelines for EU countries. The 2008 EC report is a 'Guide to Cost Benefit Analysis of Investment Projects' which contributes to shared Europe-wide project appraisal guidelines in a broad conceptual framework [8]. A more recent report was published by EC [9] named 'Guide to Cost-Benefit Analysis of Investment Projects: Economic Appraisal Tool for Cohesion Policy 2014–2020'.

These reports demonstrate an attempt at the European level to develop a common appraisal approach for major public investments.

This chapter reviews literature to identify common impacts and indicators and to examine the methodological approach concerning impact-indicator calcula- tion and valuation methods in the CBA applications of EU countries. The existing literature on these issues is limited since previous research has mainly focused on a general comparison of transport policy appraisal techniques across the EU countries and internationally [5, 10], a general review of existing tools and methods used for transport infrastructure evaluation [6, 11] or a valuation of specific indicators in a country-specific or European context [12, 13]. Less attention has been paid to specification and evaluation of the assessment methods for the priority indicators commonly used in cost-benefit assessments. To fill this gap in the literature, this chapter focuses on key impact and indicators and the techniques utilised in their assessment methodologies in a CBA framework.

2. Assessing sustainability of urban form and transport relationship

The process of globalisation and the progress achieved in telecommunication technologies have led to significant changes both in economic and spatial structure of cities [14]. The literature can be divided into two main groups: the first group points out a decline in the significance of space and distance resulting from advances in telecommunication technologies [15], while the other group supports the continuing importance of proximity and agglomeration economies [16]. Despite the existence of two different polar views in the theory, what is clear is that there is considerable migration to the metropolitan areas since the mid-1970s globally. These metropolitan centres have experienced dispersed or polycentric type of development in contrast to their monocentric structures that were common previously[1] [17]. This implies that CBD has lost its primacy and that many trip-generating activities are dispersed in clusters within peri-urban areas.

Considering its wide variety of effects on the urban area, population growth can provide benefits to the residents and the governments by attracting more special- ised services, and enhancing economic and social diversity. However, there is also a growing perception among planners and policy-makers that urban growth may not provide net gains to the society considering its negative impacts on the urban environment. An example of such an impact is uncontrolled urban growth that is associated with dispersed patterns of urban development commonly known as 'urban sprawl'. Development on the periphery, and this new form of urbanisation is seen to be problematic for several reasons: first, it necessitates providing infrastructure and services to the low-density population on the urban periphery reflecting the main cost of infrastructure provision on the society as a whole; second, it causes loss in the productive agricultural land and reduction of landscape amenity; lastly, it is related to indirect externalities such as negative effects on the air and water quality, increased travel and accessibility costs and unwanted social equity costs.

Given this framework, sustainable urban development and urban growth management have become a central issue both in planning theory and practice. Here, the main issue is the search for linkages between rural and urban spatial structure and transportation systems which will achieve sustainable urban form and efficient transport provisions (see for example [18, 19]). Efficiency in transportation which is closely related to the urban structure is generally achieved by reducing *trip lengths and times, enabling efficient transit as the dominant mode of transport, and reducing transport-related emissions, pollution and accidents* [20]. The theory suggests that compact city is preferred to more dispersed patterns in terms of sustainable spatial development and transportation efficiency [21]. The reason is related to the reduc- tion in travel demand and travel time since most of the activities are closely located in the compact form [19, 22]. It is also argued that compact form can support public transport services better than dispersed form since population densities (e.g., critical mass) in the former case are high enough to provide efficiency in different modes of public transportation[2] [20]. In the literature, there are also studies questioning the sustainability of the compact form [18, 23] and suggesting that decentralised or polycentric solutions would be better. One reason for this is that multi-centred cities provide significant transport benefits by locating residences close to employment centres [21]. Gordon et al. [24] named this as the co-location of workers and jobs.

In contrast to the previous research supporting co-location hypothesis, Gordon et al. [25] have concluded that polycentric type of development is not necessarily associated with shorter commuting distances as it is verified through the empirical work carried out for the metropolitan areas in France [26]. The results from this

[1] *Here, monocentricity is related to the single central business district, which dominated the hierarchy of business entrees by being the focal point of the transportation system while the polycentric structures are the decentralised employment centres connected by the relevant transportation networks.*

[2] *The notion of 'critical mass' is related to high population and job densities, which are interlinked with a greater demand and usage of public transport services, hence supporting the viability of public trans- port. By contrast, low densities are inefficient in terms of public transport provision, thereby creating a greater dependence on private car use.*

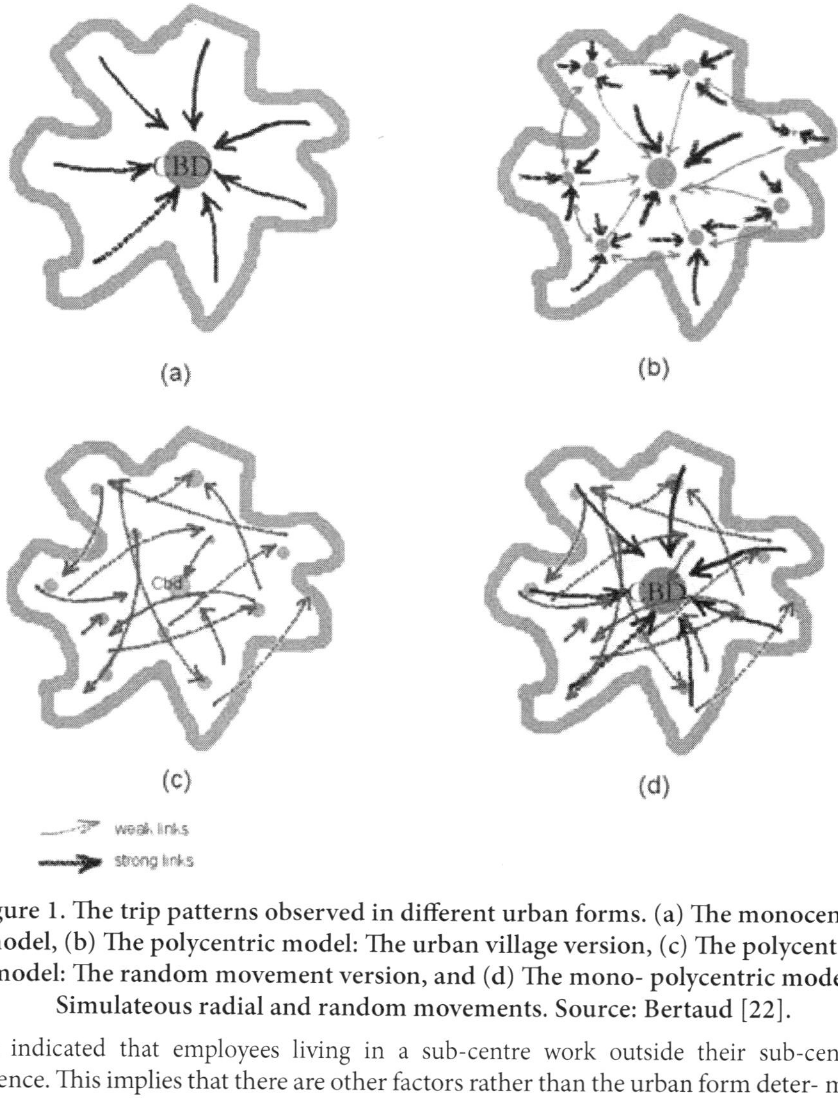

Figure 1. The trip patterns observed in different urban forms. (a) The monocentric model, (b) The polycentric model: The urban village version, (c) The polycentric model: The random movement version, and (d) The mono- polycentric model: Simulateous radial and random movements. Source: Bertaud [22].

work indicated that employees living in a sub-centre work outside their sub-centre of residence. This implies that there are other factors rather than the urban form deter- mining the commuting behaviour. The differences in commuting patterns are gener- ally explained by differences in socio-demographic factors and different preferences [27]. These factors affecting the commuting behaviour constructed a base to explain the relationship between urban form and commuting patterns. Following Van der Laan [28], four different types of commuting flows are identified: centralised,[3] decentralised,[4] exchange commuting[5] and cross commuting[6] (**Figure 1**).

The first one is the monocentric model with radial travel patterns from the periphery to the central core where a great majority of economic activity takes place

[3] *The workers live either in the suburban areas or in the City centre, and in both cases, they commute to the central city.*

[4] *Suburban areas attract commutes from other suburbs and the central city.*

[5] *People living in the central city commute to suburban areas while residents of the suburbs commute to the central city.*

[6] *There are separate labour markets in the suburban and central areas indicating that suburban residents have jobs in the suburbs while central city residents work in the central city.*

(**Figure 1a**). The existence of public transportation along radial routes results in short trip distances and travel time. The second model (**Figure 1b**) is a theoretical model of polycentricity consisting of self-sufficient 'urban villages' surrounding the CBD. In such a model, house-to-job distances are very short considering that most people are living and working in the same sub-centre, and even could go to work by walking or cycling. However, there is no example of self-sufficient sub-centres as described by this model; and therefore, the model is subject to criticism among academics (see for example Bertaud [22]). The random movement version of the polycentric model, on the other hand, could better represent the real situation as there is wide dispersion of trip origin and destinations with highly complex and random trip patterns in a multi-centred area (**Figure 1c**). The absence of radial routes and the existence of random movements, however, make it more difficult and inefficient to operate public transportation when compared with the first two models in **Figure 1(a)** and **(b)**. This stems from jobs-housing imbalances where jobs and residences are relatively far from each other and dispersed in an urban area implying a wide variety of trip origin and destinations. The last model is the mono- polycentric model (see **Figure 1d**), which refers to a multi-centric structure with a strongly dominated central core. The model provides radial movements between sub-centres and the CBD while reducing random flows between the sub-centres. In this sense, public transportation could be provided along the radial routes, but to a limited extent considering the existence of inter-suburban random flows. To sum up, these four models provide potential explanations for the relationship between urban structure and travel patterns which may exist in an urban area.

Based on these varied relationships, the concept of transport efficiency has gained increased importance. The potentially strong relationship between urban form and transportation efficiency constructed a base for the studies evaluating the economic aspects of various urban forms which will provide efficient transport systems. The literature that is related to the land use-transportation interactions can be examined under two main groups: the first group analyses the impacts of transportation networks on the structure of land development by applying methods of economic appraisal in either qualitative or quantitative framework. For instance, it evaluates the costs and benefits of transport provisions by carrying out empirical research in a given urban area in a specified time period. The examples can be found in Donaldson [29], Perl [30] and Ustaoglu et al. [31, 32]. The second group of literature derives indicators to measure and evaluate the costs and benefits of transport provisions. Under the second group, there are analyses of the appraisal of transportation projects through examining the effects on real estate property prices [33]; studies on relationships between transportation investments and land use development through deriving accessibility measures [34], and examinations of the transport-land use relationship incorporating the accessibility measures on differ- ent land development scenarios [35].

3. Transport project and policy evaluation in academic literature

The impact evaluation of transport projects and policies is to a large extent subject to technical, socio-economic, environmental and policy-based value judgements. There is no single best methodology for the evaluation of transport policy impacts; but a wide variety of methods is used depending on the nature of objectives and the characteristics of value judgements. Within the international literature, various approaches concerning transport impact evaluation can be found (for a detailed review, see [36–38]). These include economic analysis techniques such as cost-benefit analysis (CBA) [39, 40], cost effectiveness analysis

(CEA) [36, 41], and lifecycle cost analysis [42]. Additionally, transport projects can be subject to evaluation through application of specific approaches such as multi-criteria analysis (MCA) [43], social-based analysis (i.e., based on environ- mental impact assessment—EIA and strategic environmental assessment—SEA approaches) [36], decision-analysis [10] and other specific applications including land suitability analysis [44], rapid assessment methods [43], resource manage- ment approaches and simulation/ mathematical modelling [45].

Among others, CBA is the core of most assessment procedures and has an extensive use in the national appraisal frameworks of most countries in Europe and internationally (see [46, 47]). As the OECD suggests, the CBA assessment can consider a large set of criteria including financial impacts for transport providers, effects on transport user benefits, socio-economic impacts, transport network effects, energy and environmental impacts and policy impacts beyond the trans- portation system [48] (see also [49]). As a result of the difficulties in quantifying some of the impacts in monetary terms and the existence of objectives which are not always related to economic efficiency, CBA may not be a desired option for project evaluation in every case [50, 51]. In order to address some of these issues, other approaches outlined above have been introduced and applied internationally either as modified alternatives to CBA or in a complementary analysis frame- work [10, 36, 38].

Related to the transportation evaluation methodologies provided in the literature, Litman [52] points to some of the weaknesses of transport-land use impact evaluations and suggests using a more comprehensive approach for evaluating the impacts of transportation provisions on land use, travel activity, land development patterns, and land use accessibility and transport diversity relationship. This has been initially highlighted by many studies in the literature (see [48, 53, 54]) indicating that a wider scope for the assessment of transportation projects is increasingly needed to cover externalities of transportation investments which were previously not included in the traditional cost-benefit assessment methodologies.

Additionally, there is further research criticising standard cost-benefit appraisal techniques as they do not consider wider economic benefits of transport investments [55]. One of the key recent studies carried out in this area is that of Graham [56], who investigated external agglomeration benefits arising from the provision of transport infrastructure. This study is based on Venables' [57] earlier research regarding the impacts of transport investments on agglomeration of industries and further impacts on the economy. In this respect, agglomeration effects are crucial for both manufacturing and service industries as they create connections between firms for achieving benefits from labour market pooling, the sharing of intermediate inputs, and knowledge sharing or technological spillovers

[58]. Graham's [56] research points out some positive externalities from increas- ing urban densities associated with a transport investment supporting Venables' argument concerning the significance of agglomeration externalities in urban economies.

Contrary to these arguments highlighting beneficiary impacts of transporta- tion investment on economic growth, literature is unclear on the degree to which economic development benefits stem from the transportation investments (see for example, [59]). The general argument in the literature is that benefits from economic growth are mainly represented in travel cost savings which result from improvements in the efficiency of the transportation system. Travel cost savings include the savings in travel time, vehicle operation costs and costs of accidents, reduction in traffic congestion, etc. Some studies argue that inclusion of economic growth effects will lead to double counting since 'economic growth benefits are the manifestation of capitalised travel cost savings' ([48], p. 19). Despite a lack of consensus on transport infrastructure investment and economic development relationship, there is an increasing number of studies in the literature focusing on development and assessment of wider economic impacts in transport project and policy appraisals (see [49, 60]).

In addition to the wider economic benefits, some other impacts of transport proposals were specified by Nash and Preston [61] aiming at a more comprehensive assessment in a conventional CBA approach. They argued that financial appraisal frameworks of transport investment programmes should be extended to cover external benefits including traffic congestion, environmental and developmental benefits and others. Related to developmental benefits, empirical evidence has shown that development of a transportation system results in changes in property values along the transport corridors [62]. These studies point out that the results indicate the short-term impacts of the development of a new transport system rather than the long-term market effects. The degree to which increases in property values stem from transport investments is unclear due to the existence of other forces influencing real property markets. These may include the changing planning and policy settings within the area, regional economic development trends and forces, and the availability of other transit connections and developable land within the area [63]. It is also important to mention the existence of relocation effects stemming from local development, that is, the gain achieved by one area may be lost in another area in the region implying a net zero effect overall. Based on these issues, property values, on general basis, are evaluated separately in the transportation impact evaluations on land development.

Many studies in the literature have argued that transportation systems can play a critical role in strengthening or weakening the problems posed by the dispersed or sprawl type development patterns [52]. Dispersed development necessitates providing public services to the low-density population; causes loss in agricultural land and reduction of landscape amenity. It is also related to indirect externalities such as: negative effects on the air and water quality, increased travel and accessibility costs. In order to represent land development impacts of transport provisions, there are a group of indicators suggested to be included (e.g., public service provision costs) for the cost-benefit evaluation of transport investment assessments. The details of the indicators can be seen in Litman [52].

Concerning environmental impacts, transportation system, particularly road transport is accepted to be a main contributor to the increasing levels of green- house gas emissions

[64]. The dramatic increase in private vehicle ownership, which is also encouraged by the provision of large-scale urban motorways, has led to air/noise pollution and increasing amounts of transport-related energy consumption. In contrast, the average rail and transit passenger tends to produce less carbon dioxide than a road transport user [65]. Although there are examples of counterarguments, the general research has been in favour of compact urban form in comparison to the more dispersed urban development largely on the grounds of transportation energy savings (see [18]). Energy and environmental impacts, that is, energy consumption, air/noise pollution exposure, climate change emissions (greenhouse gas emissions) are all important for the transportation evaluation process and are generally covered internationally in transport project assessments.

Alternatively, the study of Nash and Preston [61] points out some possible changes in other public transport revenues (e.g., bus) following a shift from existing public transport use to the newly introduced public transport network (e.g., rail). Such changes in public transport revenues need to be included in evaluations as a newly introduced public transport infrastructure results in revenue losses for the existing public transport operators resulting from a reduction in demand and consequent reductions in services and costs to compensate (see [61]). Another issue is the changes in tax revenues resulting from shift of demand among various modes of transportation users. As an example, introduction of a new rail system can have influences on car tax revenues possibly implying a loss in revenues from taxed road transport since a number of road users may shift to untaxed rail.

Another issue which is of great concern in benefit-cost evaluations of trans- port investments is the assessment of public funds associated with a project proposal [8]. Projects have impacts on public funds through the need to finance capital expenditures and the impact of the project on taxation receipts [66]. The tax revenues generated by the project (i.e., direct and indirect tax revenues) can be evaluated in two ways: either they may decrease the need to finance budgetary deficits by public debt or taxation, or they provide the opportunity to increase public expenditure [66, 67]. Estimation of marginal benefits of additional public expenditures is cumbersome; therefore, it is suggested to apply the marginal costs imposed on the economy by the collection of additional public revenues [66, 67]. One of the main sources of public revenues is tax collection; consequently, the marginal cost of public funds can be calculated as the cost to the economy of collecting an additional unit of tax revenue (see [67]). From the perspective of project finance, a shadow price of public funds exceeding a unitary value implies that each €1 raised through taxation gives rise to a social cost in excess of €1. The EC Final Report [8], p. 57, defines the marginal cost of public funds as 'the ratio between the shadow price of tax revenues and the population average of the social marginal utility of income'. These are country-based values depending on the taxation system and suggested to be used to adjust the flows of public funds to and from the project.

In line with this literature, the key impacts and indicators in the evalua- tion of transportation-land use relationship are summarised in **Table 1** with the examples of studies particularly focusing on these specific impacts in their detailed research. The literature may also cover some other impacts and indica- tors to be included in transport policy and project evaluations. As these are not fully covered in this paper, their details can be seen in related literature including Litman and Burwell [98]; Marsden et al. [99]; Sinha and Labi [38]; UN [100]; and Litman [97]. The comprehensive approach summarised in **Table 1** can be consid- ered as an enhanced approach in evaluating the impacts of transportation policy and provisions. The variety of impacts given under this method implies that there are various interest groups which will be affected from the development and implementation of transportation policies in different ways. Janic [96] identifies these groups such as: users; systems' operators; public, semi-public and private investors; policy-makers at local (regional) and national level; and local commu- nity members (see [96], p. 496).

The policies and their impacts—namely costs and benefits—can be conflict- ing for particular groups. For example, new transportation investments such as rapid transit systems will benefit direct users by reducing time and money spent for transportation; operators by increasing their profitability; investors of transportation infrastructure by increasing their rate of return, policy-makers by improving the efficiency of the transport sector and the economy; and com- munity members by improving their local socio-economic conditions. On the other hand, transport investments can imply different costs to different interest groups: travel costs may increase for the direct users whereas it may become costly for operators to run a high-tech business. Investors and policy-makers may come up with high infrastructure costs while community members may suffer from local impacts on their environment. It is clear from this example that implementation of new transport policies represents a trade-off between gains and losses within and between different interest groups. Therefore, it is important to consider these trade-offs by including as many of the policy impacts into the transport project evaluations. Accordingly, Litman [52] points out to some of the weaknesses of current project evaluations and suggests using a more comprehensive approach for evaluating the land use impacts of trans- portation provisions [52]. The main steps in the evaluation of transportation- land use relationship are summarised in Table 2.

Table 1. **Summary of impacts and indicators for the transport infrastructure evaluation.**

Impacts/indicators*

1. Direct impacts of transportation infrastructure provision:

- *Transportation facility land values:* The cost of land used for transportation infrastructure construction and other public facilities dedicated for transport vehicle use [52, 57]
- *Infrastructure development and construction costs:* The cost of designing and constructing transportation facilities including land and transport infrastructure construction [68]
- *Traffic services:* These refer to the costs of police, emergency response, law enforcement, planning, street lighting, parking enforcement and driver training [52]
- *Adjacent property values:* The change in real property values resulting from the provision and operation of the new transportation infrastructure [69]

2. Socio-economic impacts:

a. Land development impacts:

- *Green space preservation: Refers to effects of transportation activities and facilities on the green space, e.g., parks, gardens, farms, woodlands, etc. [52]*
- *Public service costs: These point to how costs of public service provision tend to increase with dispersion of urban activities [70]*
- *Urban sprawl: Land development impacts vary by mode since private car-based transport requires more space than other modes for travel and parking, and tends to encourage more dis- persed patterns of land use. By contrast, alternative modes such as bus transit and rail systems are more likely to contribute to more compact and mixed-use land developments [71]*
- *Regeneration: The provision of a new transport system associated with corresponding land use plans and policies can be influential in promoting urban renewal particularly in unfavourable urban areas [72]*

b. Transportation-related impacts:

- *Safety: The ability of the transportation system to allow users to move freely without damage and harm [73]*
- *Vehicle ownership and operation costs: Direct user expenses for the ownership and use of private vehicles [74]*
- *Transit fares: Costs and revenues of public transport fares to the users and system providers [75]*
- *Travel time: Time spent on transportation including waiting and actual travel [76]*
- *Comfort and convenience: This refers to the quality of transport service including ride quality, crowding and quality of information, cleanliness and ambience [77]*
- *Traffic congestion effects: Refers to incremental delay, vehicle operating costs, transport-related pollution and stress resulting from interaction among vehicles in the traffic [13]*
- *Transport diversity: Quantity and quality of travel options (particularly of non-drivers') are considered [78]*
- *Barrier effects: Delays, discomfort, lack of access that roads and traffic cause to non-motorised travel [79]*

c. Socio-economic development impacts:

- *Affordability (housing): Potential expansion of accommodation choices for all individuals to increase mobility and lower the combined cost of housing and transportation [80]*
- *Affordability (transport): People's ability to access basic goods and activities (housing, medical care, food, education, work and social facilities) by means of transportation [81]*
- *Social inclusion: Transport-related factors influencing individual's ability to access education, employment and public services, social and recreational activities [12]*
- *Socio-economic growth: The development and growth impacts of transportation infrastructure on the economy and the society [82]*
- *Wider economic impacts: Introduction of a local transport investment is influential in changing the effective density of employment and jobs that are accessible to the local economy. This will have further impacts on productivity and efficiency, i.e., agglomeration externalities, competi- tion effects, output effects (of imperfect competition) and labour market effects [83]*
- *Land use/transport accessibility: The ability of the transportation system to connect people to goods, services and activities, and to meet needs of different populations [84]*
- *Area property values: Transportation policies and planning decisions have influences on property values as well as the location and type of real property development [62]*

Impacts/indicators*

d. Impacts on government fiscal balances

- *Changes in tax revenues: These represent changes in transport-based tax revenues following a demand shift among different transport modes [85]*
- *Marginal costs of public funds: Refers to the impacts of transport projects on public funds through the need to finance capital expenditures and the impact of the project on taxation receipts [86]*

3. Transport network effects:

- *Reliability: Variation and consistency in travel times and the reliability related to external factors [87]*
- *Quality of transport service: Relates to ride quality, crowding, ambience and quality of information [88]*
- *System operating and maintenance costs: Refer to expenditures to maintain the transport facilities including maintenance and operations [89]*

4. Energy and environmental impacts:

- *Climate change emissions: Refers to the greenhouse gases (i.e., CO2, NOx, CH4) emitted from transportation vehicles and related facilities that increase atmospheric solar heat again [90]*
- *Air/noise pollution exposure: The noise and air pollution associated with transportation system construction/operation [91]*
- *Resource consumption costs: These refer to various direct and indirect costs of energy produced, distributed and used in vehicle and transport facility construction and operation [92]*
- *Water pollution: Pollution (surface and ground water) associated with transportation facilities and vehicle use [93]*
- *Waste disposal: External costs resulting from vehicle waste disposal activities [94]*
- *Ecological impacts: Transport infrastructure and operation impacts on flora, fauna and their habitat such as wetland [38]*
- *Landscape and heritage: Transport networks and related facilities, vehicle traffic and low-density development can be a threat to cultural heritage and often degrade landscape beauty [95]*

It is obvious from **Table 2** that transportation provision has various impacts on land use; and as empirical evidence has shown, it is difficult to quantify most of these impacts. Some of the impacts and indicators can be represented in monetary values while the others can only be expressed in a more qualitative way. Here, an important issue to consider is that there may be correlations among various indicators such as the positive correlation between land use accessibility and land values, or the negative correlation between air pollution exposure and area property values. Considering the correlation effects, indicators should be kept as orthogonal as possible in order to prevent the double counting problem in the transport policy evaluations. Therefore, selection and confirmation of the most relevant indicators is an important stage within the project evaluation process.

Transport provision results in:	Physical effects	Impacts	Examples of indicators	Expected direction of change for stakeholders
Direct changes in land usage	Amount of land devoted for transportation facilities	1.Green space preservation	1.Per capita green area used for transportation facilities (QN)	Down: P, C
		2. Transportation facility land values	2. Estimated value of the land used for transportation facilities (QN)	Down: I, P
		3.Development costs/ capital investments	3. Costs of demolition, levelling, reinforcement, etc. of land (QN)	Down: I, P
		4.Adjacent property values	4. Estimated value of the adjacent properties (QN/QL)	Up: LC, P
Changes in development patterns	Location, density and compactness of development	5. Green space preservation	5. Per capita impervious surface land (QN/QL)	Down: P, C
		6. Public service costs	6. Costs of providing public services such as roads, utilities, etc. (QN/QL)	Down: P, C

Changes in land use accessibility and transport diversity	Dispersion of common destinations, and quality of travel options	7. Changes in per capita vehicle travel	7. Changes in (QN/QL):	
			7.a. vehicle operation costs	Down: U, C
			7.b. travel time	Down: U, C
			7.c. risk of accidents	Down: P, C
			7.d. comfort and convenience	Up: U, C
			7.e. traffic congestion	Down: P, C
		8. Area property values	8. Estimated value of the properties within area (QN)	Up: P, C
		9. Socio-economic benefits		
		9.a. Affordability (housing)	9.a. Affordable housing accessibility	Up: P, C
		9.b. Affordability (transport)	9.b. Portion of households' budgets needed to provide adequate transport (QN)	Down: P, C
		9.c. Social inclusion	9.c. Employment/income/ education, etc. levels (QN/ QL)	Up: P, C
		9.d. Socio-economic growth	9.d. Growth in output, employment, etc. (QN)	Up: P, C

Table 2. Impacts of transport provision on land use.

Transport provision results in:	Physical effects	Impacts	Examples of indicators	Expected direction of change for stakeholders
Changes in land use accessibility and transport diversity	Dispersion of common destinations, and quality of travel options	10. Land use/ transportation accessibility	10. Quality of land use/ transportation network accessibility (QN/QL)	Up: P, C
		11. Transport network effects		
			11.a. Reliability/quality of transport service (QN/QL)	Up: O, P
			11.b. Systems' operating cost (QN)	Down: O, P
Changes in travel activity	Per capita motor vehicle ownership and use	12. Consumer transport costs	12. Generalised travel cost (QN)	Down: P, C
		13. Accidents	13. Crash death and injuries/ related economic costs (QN)	Down: P, C
		14. Energy and environmental impacts		
		14.a. Energy consumption	14.a. Total vehicle emissions (QN)	Down: P, I

		14.b. Air/ noise pollution exposure	14.b. Ambient air quality/ noise levels (QN/ QL)	Down: P, C
		14.c. Climate change	14.c. Climate change emissions (CO2, CH4) (QN/QL)	Down: P, C
		14.d. Water pollution		Down: P, C

Note: P: policy-makers; C: wider community members; I: investors; LC: local community members; U: users; O: operators; QN: quantitative; QL: qualitative. Source: Adopted from: Janic [96]; Litman [52].

4. Key base indicators for the cost-benefit applications

Based on the literature focusing on transport project and policy appraisals, this section reviews several key impacts and indicators which can be utilised for CBA applications in European countries. Scenario analysis—comprising a baseline scenario (reference scenario) which is compared with several alternative scenarios— can construct a base for the impact evaluation of public transport investments (e.g., rail-based transport). As a priority, two alternative scenarios can be defined: a baseline *business-as-usual* and alternative *with rail* scenarios. According to the baseline scenario, it is assumed that the urban area would continue to grow with the present trends and there would be only sufficient maintenance and renewal investments to maintain the existing infrastructure. Therefore, rail services in future years would be broadly comparable to the current level potentially leading to a more dispersed urban form. Considering the land development impacts of rapid rail investments, it can be expected that the *rail* scenario will generate more compact forms of urban development. By encouraging a transfer from private transport, rapid rail investments can assist in improving accessibility and land use change which supports compact and mixed developments (see Litman [52]).

The potential for efficiency and environmental impacts can be examined through the impacts and indicators specified in the previous sections. The economic appraisal process can be summarised in six stages (see **Table 3**). In accordance with the estimations from the transportation model specified for the study and the parameters/ values specified for the capital costs, costs of accident, vehicle (and system) operation, public service provision, travel time and global and local air pollution, **Table 4** presents the related data requirements for the scenario analysis of baseline and *with rail* cases. Based on the impact evaluation data given in **Table 4**, some specific issues in impact-indicator valuation methods are then explained in the following sections.

4.1 Capital cost estimation of transport investments

A broad estimate of the capital costs for any public transport provisions is nor- mally obtained at project initiation stage. These estimates are expressed in constant prices and are generally built up using unit cost data, expert advice and experience of similar projects in the past. Given the inherent uncertainty at this stage, a detailed risk analysis is necessary to reduce uncertainty around the expected infrastructure costs of transport projects. In the literature, there are examples showing that cost escalations are common in transport infrastructure projects, particularly in urban rail projects. Flyvbjerg et al. [101] showed that transport infrastructure projects worldwide experience large construction cost escalations; and among them, rail projects incur the highest cost escalation. The average cost escalation for rail projects is 44.7%, followed by fixed-link projects (bridges and tunnels) by 33.8% and road projects by 20.4% [101]. Practical methods for risk assessment and management in urban rail projects are provided in Flyvbjerg and COWI [102] where UK transport projects were grouped and evaluated. This research developed capital expenditure up-lifts for urban rail projects in the UK. Other studies in the US, for instance, have also found that the mean overrun of light rail transport projects is around 42% [103] while the mean value is more than 20% regarding rail projects in the US [104].

Table 3. Stages of economic appraisal process for public transport investments.

1. Forecasting transportation demand with a transportation model consisting of:

- Forecasts of future growth and land use change (population, employment, economic activity, income)
- Assumptions of the supply side of transportation activities
- Assumptions and scenarios for external conditions
- Four-stage method: trip generation, trip distribution, modal split, network assignment

2. Quantifying, where possible, incremental costs and benefits relative to the baseline scenario
3. Identifying unquantifiable impacts
4. Adjusting quantified costs and benefits for:
    - Inflation
    - Relative price changes
    - Risk and optimism bias
    - Undertaking sensitivity analysis
    - Calculating the net present value
    - Adapted from: EC Final Report [8].

Table 4. Impact evaluation data for rail-based infrastructure investments: with rail vs. a baseline scenario approach.

Impacts/indicators	**Impact evaluation data requirements of *with rail* vs. baseline scenario**
1. Capital costs of transport infrastructure investment.	Direct construction cost estimates include the following: land acquisition costs, railway infrastructure, stations, civil engineering works, operational systems, planning and design
2. Provision of public services	Future estimated numbers for new residential development (numbers of new housing units) in the case study area within the appraisal period specified for rapid rail investments
3. Accident rates/future accident risks	Three types of data are required: • The most recent data related to the number of personal fatalities, serious injury, and minor injury accidents along the catchment area of the newly proposed transport line • Estimated numbers for future accident risks from the national and local accident rates and trends • Quantification of changes in the number of fatalities, serious injuries, and slight injury accidents due to a rapid rail investment by using country specific risk functions

4. Change in road vehicle operation costs	For the calculation of the economic benefits (costs) associated with vehicle operating costs, two types of data are required: • *Demand: the number of private vehicles (cars) making a particular origin-destination trip for the baseline scenario and the alternative with rail scenario (peak/off-peak traffic flow data for the baseline and alternative scenarios)* • *Vehicle kilometres: total change in vehicle kilometres from the local highway network for the baseline and with rail cases*
5. Change in travel time	Estimates related to: • Travel time-change in travel time for private vehicles (cars) in peak/ off-peak traffic for the baseline and with rail scenarios • Demand: peak/off-peak traffic flow data for the baseline and with rail cases
6. Public transport operating costs and revenues	• Expected operating pattern and service frequency of newly proposed rapid rail system • Key characteristics (route length, journey time, peak and off-peak headway, etc.)
7. Change in emissions	• Total change in greenhouse gas emissions (i.e., CO2, in particular) and • local air pollutants for the baseline and with rail cases.

also found that the mean overrun of light rail transport projects is around 42% [103] while the mean value is more than 20% regarding rail projects in the US [104].

Based on this research, adjustments for capital cost bias should be added to the initial cost estimations in the countries where public transport projects (e.g., rail- based transport) are being constructed. However, a detailed risk analysis can also reduce the uncertainty in cost estimations, and therefore, the need for bias adjustments can be reduced. For EU countries other than the UK, HEATCO [105] (a EC 6th Framework Programme) suggests an average of 34% capital expenditure up-lift on original cost estimates for the rail projects based on the results in Flyvbjerg et al. [101] representing average cost escalations in Europe.

4.2. Valuation of traffic safety

It can be argued that any limited economic analysis underestimates the value of human life to family and society more generally. However, for the purpose of CBA evaluation of transport projects, this must be considered even if the approaches may be limited and contested. In the literature [105–107], some measure of the statistical value of human life (SVHL) is often used and has been determined using three methods: cost of restitution, human capital and stated preference approaches. The first method represents the direct costs generated by accidents. Human capital approach measures discounted loss of production due to injury or death of the individual member of the workforce while the last method is used for estimating willingness-to-pay (WTP) values of individuals indicating their preferences to reduce the risk of being injured or dying in an accident. The literature examining the methods for estimating the statistical value of life is vast. Some examples are Viscusi and Aldy [108]; Treich [109] and Woods et al. [110]. This literature critically reviews the existing methods utilised in a specific country or cross-country comparisons which were undertaken based on safety valuation considerations. It is shown that there is no single method used to determine the statistical value of human life, but a variety of methods have been used by different countries.

The literature points towards the WTP approach as being a widely used methodology in the EU. Considering this, Grant-Muller et al. [46] identify some factors which may contribute to the variations of WTP values across EU countries. These include: (1) income (per capita) variations among member countries influencing an individual's WTP values for safety; (2) cultural differences which have an impact on government attitudes and individual tastes and preferences for accident reduction measures; (3) inclusion or exclusion of some costs, that is, legal costs, and other public sector costs and (4) differences in the nature of the measurement methods used, that is, problems of bias in WTP measures. Based on the existence of these factors, HEATCO [105] suggests using the human capital measures and WTP studies carried out in the countries for which they are applied. Based on the existence of these factors, HEATCO [105] suggests using the human capital measures and WTP studies carried out in the countries for which they are applied. Considering the country-specific differences, another EC research project—UNITE—also sets recommendations which allow adjustments to EU countries of a common European set of values. EC [111] has provided updated accident costs for the EU countries which were initially provided by HEATCO [105] (**Table 5**).

Increasing values for future years is based on the estimation of a country- specific rate of growth in real GNP (or GDP) per person employed. HEATCO [105] recommended a default inter-temporal elasticity to GDP per capita growth of 1.0. In contrast to cross-sectional elasticity, inter-temporal elasticity to GDP considers underlying changes in individual preferences and technology over time. Further to this, sensitivity testing with an income elasticity of 0.7 is suggested by HEATCO [105] if accident costs contribute a significant part of benefits in cost-benefit assessments. Bowland and Beghin [112] is an example of a meta-analysis based on contingent valuation studies examined in the literature. Their results show statistically significant income elasticities of 1.7 and 2.3. Viscusi and Aldy [108] carried out a comprehensive meta-analysis for the value of statistical life estimates throughout the world. Their results point to estimates of the income elasticity in the range of 0.5–0.6. Considering that the literature on the income elasticity of value of statistical life is controversial, country-specific values are recommended for the value of statistical life analysis. The cross-country differences in socio-economic factors, particularly variations in income, and the type of safety projects considered in a specific country can show considerable variations in the value for assessing risk. This raises the issue of development of a specific methodology by each country through utilisation of the related data and application of value of statistical life analysis.

Table 5. Examples of EU values per casualty avoided (2010, purchasing power parity, market prices).

		Accident type	
	Fatality (Million €)	**Serious injury (Million €)**	**Slight injury (Million €)**
France	2.070	0.2892	0.0216
Germany	2.220	0.3071	0.0248

Ireland	2.412	0.3056	0.0233
Italy	1.916	0.2462	0.0188
The Netherlands	2.388	0.3164	0.0255
Poland	1.168	0.1567	0.0113

Source: (see) EC [111] for all EU country values.

4.3. Vehicle operation costs

4.3.1. Road vehicle operation costs

Vehicle operating cost savings are associated with user benefits indicating the shift of travel from private car to public transit. At a minimum, the shift from private car to public transport systems saves fuel and oil, which can be considered to have important impacts on energy consumption and environment pollution levels. In addition, there are costs of depreciation, insurance and parking which are affected from increasing car usage in the way that there are increases in repair and maintenance costs, reductions in vehicle resale value, increases in parking and traffic costs among others [113]. The unit vehicle operating costs are clearly dependent on the prices of goods within a region (i.e., price of oil, vehicle parts, etc.), the transport network characteristics and vehicle utilisation. However, operating cost relationships for road vehicles are more generic and transferable between countries [105]. There are generic models and computer software for the calculation of road vehicle operation costs in the absence of a local model. The Highway Design and Maintenance Standards Model (HDM) is an example of such a model which is recommended by HEATCO [105] and World Bank for both European and World Bank-funded transport projects. The HDM, developed by the World Bank, has been used for over two decades for the assessment of road investment programmes and analysis of road network strategies in many countries [114].

In the HDM, the costs to road users for a given country are represented as a func- tion of vehicle fleet unit costs, utilisation and characteristics, and road characteristics [114]. The model provides annual estimates (for a given road strategy) such as the road condition and resources used for maintenance, the vehicle speeds, physical resource consumption of vehicles, individual vehicle operation cost components and total vehicle operation costs. Either consisting of all available road network data or use the road network aggregate data if detailed network data is unavailable. From this data set, a representative matrix of road classes is developed and utilised for the estimation of road vehicle operation costs. A considerable effort is needed to quantify the attributes of a road class, that is, the length, width, pavement type, climate zone type, geometry type, traffic composition, roughness, surface condition type, drain type and construction quality [114].

The road attributes, mostly based on aggregate data, can only be estimated through an expert judgement or can be obtained through statistical analysis of the data available in the road network database [114]. A further issue arises with the estimation of the network length when traffic load, road condition data and other road attributes are not collected by the same road agency unit or at the same time [114]. These are some issues of concern about the difficulties of the HDM applications in estimating an operating cost relationship for road vehicles. Following the estimation of parameters in maintenance and vehicle operation, unit costs and specified prices are applied to determine the maintenance and vehicle operation costs. In this respect, local relationships and prices are strictly recommended to be used for the calculation of vehicle operation costs [105].

HDM can be considered as a very useful tool for calculating the vehicle operation costs. HDM-III and HDM-4 are the later versions of the software which have been improved by adding new models for road deterioration and for operating costs of numerous vehicle types. These models predict the change in road roughness with respect to cumulative axle loads or maintenance actions by the road administration [115]. The estimation of vehicle operation costs requires both local data on vehicle ownership and repair costs and a generic relationship between road roughness and vehicle operating costs. This implies that both local models and default valued models are available and can be utilised in the HDM [114]. Some recent examples on the applications of HDM are Cutura et al. [116] and Perrotta et al. [117].

4.3.2. Public transport (e.g., rail) operating costs

Railway costs can be analysed as fixed and variable costs: fixed costs are incurred costs for operation, maintenance and replacement, which are independent of traffic volume changes. Variable costs, on the other hand, are those which depend on traffic volume. The elements of the costs for railways are specified by the World Bank [118] Infrastructure Reports (see Table 6). Unlike road vehicle operation costs, rail operating costs can be influenced by regulatory and institutional characteristics of the countries in which the rail system operates. Furthermore, physical characteristics of the rail network (depot locations, track alignment, etc.), operational characteristics and labour market conditions are all effectual in determining the operating costs [118]. Therefore, country-specific data and local relationships on rail operating costs will be the most appropriate to be utilised in operating cost estimations. The availability of such data will depend on the accounting practices within rail transport sector of each specific country.

It can be suggested that cost items given in Table 4 can be calculated by directly assigning actual expenses to actual operations or through revenue analysis where revenue expenditures have changed historically with train operations [119]. Additionally, cross-sectional analysis of revenue accounts by train operations can be recommended if there is geographical variation of accounts [118]. However, these analyses may not be appropriate in a situation where the cost base will be affected by the proposed transport investment. Examples for the latter case include ([118], p. 6): (a) the use of a new locomotive with unknown operating costs and reliability; (b) a considerable change in the level of service provision at the regional level (e.g., congestion effects may result in a requirement of new infrastructure with different utilisation rates) and (c) major policy reforms including privatisation or commercialisation of railways.

Table 6. Elements of rail operating costs.

Cost type	
Vehicle ownership costs	
Locomotives/coaches	Replacement cost
Vehicle maintenance costs	

Locomotives/coaches	Unit cost/loco. Unit-km
	Unit cost/coach-km
	Unit cost/coach-year
Transportation costs	
Train fuel	Unit cost (gross tonne-km)
Train crew wages	Actual by cost centre
Locomotive crew wages	Actual by cost centre
Station operations	Unit cost/train-km
Billing	Unit cost/car load
Other	Unit cost/train-km
Source: Anderson [119], World Bank [118].	

4.4. Value of time

There are two categories of time involved for valuing travel time savings of passengers including working and non-working trips. The former is related to commuting for working purposes while the latter comprises all other non-working activities such as retail and leisure. Travel time is evaluated by standard values of time for each vehicle category assuming a constant marginal unit value of time regardless of the time saved and the variance of income levels of individuals. The valuation is based on three sources: first, a cost-saving approach, which considers wage rates as a measure of productivity loss or gain by the labour force, is selected as a minimum approach for the valuation of work time savings [105]. Second, an alternative methodology proposed by Hensher [120] identifies work trips having two components: a business component (which assumes that not all travel time is unproductive) and a private component (assuming not all savings are transferred to extra work but any utilised for non-work purposes). An alternative approach is based on the idea of willingness-to-pay (WTP) which is used for the valuation of all non-work trips and the private component of work trips. Stated preference and revealed preference are the two methods used in WTP analyses for the purpose of generating a differentiated structure of values of time (for work and non-work activities).

The cost-saving approach is thus criticised as approach assumes no utility impact on transport users and that all travel time savings can be transferred to productive output [105]. Hensher's [120] approach, on the other hand, provides a more sophisticated assessment of work time savings by assuming that part of travel time is productive and there are utility impacts on workers as not all savings are transferred to extra work. As the latter approach is more comprehensive, such an approach can be preferred in the valuation of work time savings in the EU. However, the complexity of the analysis and data accessibility problems in some European countries limits the application of the Hensher's approach. A cost-saving approach is more practical; and there is evidence in the literature suggesting that the cost-saving approach is a reasonable approximation to the social value of work travel time savings [121]. Mackie et al. [122] claim that a cost-saving approach is reasonable for the estimation of commercial value of travel time. However, an appropriate estimation of social value requires an assumption of full employment of the relevant class of labour. In the case of widespread unemployment, there will be divergence between the com- mercial value of time savings to firms and the social value of time saving. Mackie et al. [122] recommend the use of a shadow price regarding this situation.

Given the above literature, the use of national values can be suggested for net average hourly wages for work time valuation, and for the non-work time valuation use, the values in national value of time studies published in the countries of interest. There are various countries that have already published national value of time studies including the Netherlands, Norway, Sweden, Finland and the UK [123]. There are issues of differences in the WTP analysis conducted in different countries. Mackie et al. [122] identified six key factors influencing individual's value of travel time savings. These include: journey time, characteristics of the journey (congested, free-flow or repetitive), journey purpose, journey length, mode of travel and size of the time saving. Correlations with personal characteristics are also required to be introduced to separate such impacts from the others. This approach is highly data intensive and a certain level of disaggregation is needed to carry out the choice experiments specified for any country. As an attempt to set a common guideline for the estimation of travel time savings in the EU countries, HEATCO [105] and Wardman et al. [124] identified a minimum disaggregation level for the calculation of travel time savings as well as the estimations which require local data or outcomes of research which can be transferred from some other sources. These studies specified national values for travel time savings by utilising meta-analysis models to estimate travel time values for each EU country. These are relevant for use in countries where there are no data and research conducted in the field of the value of travel time savings.

4.5. Environmental impacts

Global air pollution which is mainly caused by carbon emissions can be calculated by utilising the social cost of carbon approach. The social cost of carbon can be defined as the welfare loss due to an increase in carbon emissions [125]. Due to the uncertainty of future emissions and climate change, there is wide uncertainty among the SCC estimates. Kuik et al.'s [126] meta-analysis study is well known in the literature verifying this considerable variability across the SCC estimates (see also [127]). Following a review of some key studies on climate change, Tol [128] found a median estimate of €4 and a mean of €25 per tonne of carbon emitted.

However, these estimates are conservative due to the fact that only damage which can be estimated with a reasonable certainty is included in the analysis and impacts such as extended floods and frequent hurricanes with higher energy density are excluded as there are data limitations explaining the possible relationships between global warming and these impacts (HEATCO D5, [105]).

Alternatively, HEATCO points to the research performed by Watkiss et al. [129] on the social costs of carbon. In this study, shadow price values for carbon are derived considering the future development expectations of damage and abatement costs of carbon. Damage costs were estimated on a general basis whereas abatement cost estimates are based on the UK Government's long-term goal of meeting a 60% CO_2 reduction in 2050. It is noted that the latter is consistent with the EU's target of limiting global warming to an increase of 2°C of the earth's average temperature above pre-industrial levels (HEATCO D5, [105]). This study is the most comprehensive exercise among others confirming the assumption that emissions in future years will have greater total impacts than emissions today, thereby stressing future increases in value estimates.

Local air pollution caused by road transportation is highly case specific since it has impacts on human health and the environment in local areas. The main pollutants (i.e., NO_2, SO_2, NMVOC, PM_{10} and $PM_{2.5}$) are directly related to the number of vehicles travelling on local roads, and therefore, the change in number of vehicles results in changes in concentrations of emissions in the affected areas. The country- specific estimated values which can be used in the absence of such research in any EU country were derived by using the methodology developed in the EU projects.

The Impact Pathway Approach (IPA) is an example of the methodology which was developed for the environmental impact evaluation in the ExternE project series [130]. According to the IPA approach, a transport activity causes changes in concentrations of air pollutants, which have impacts on various receptors including human beings, materials or ecosystems. This results in direct and indirect impacts on the utility of affected persons. The valuation of these welfare changes follows the willingness-to-pay (WTP) or willingness-to-accept (WTA) approaches for the damages to human health. For materials or crops, market prices can be used directly to evaluate the damages. Concerning the health impacts, three components of welfare changes have been identified [130]. These are: (1) resource costs (i.e., medical costs); (2) opportunity costs (i.e., productivity losses) and (3) disutility. The first two components named as 'cost of illness' can be estimated by using market prices which are particularly case specific and require local data. The last component, on the other hand, implies a higher value than the cost of illness [105] and requires specific modelling to represent individual's loss of welfare.

In the ExternE projects (e.g., NEEDS project), country-specific impact and cost factors were calculated using the EcoSense software tool and impacts and resulting costs occurring in Europe were calculated for increasing the existing emissions of NOx, SO_2, $PM_{2.5}$ and NMVOC by 10% in each country [105, 131] (**Table 7**).

Impacts and costs were compared to those calculated for the reference scenario, which implies that the difference between both scenarios is caused by the additional emissions. This methodology can be criticised on the basis that the estimates may not accurately represent local population density exposure and national vehicle fleet compositions. Therefore, a detailed case-specific exposure modelling can be provided where possible, taking into account the pollutant dispersion modelling and estimation of changes in the population's exposure to the related pollutants [105]. Concerning local air pollution assessment, trends in air pollution exposure are difficult to establish as software packages to measure repeated exposure are not yet available. Therefore, population-level estimates do not adequately represent extreme individual exposures [132]. There are differences in exposures since some people experience heavy traffic and influenced by higher level of pollutants than others. In addition to this, exposure to transport-related air pollution is difficult to separate from exposure to total air pollution [132]. These are some key difficulties in relation to developing a common methodology in the assessment of local air pollution across the EU countries. Country- and case-specific modelling is therefore essential for an appropriate estimation of impacts and resulting costs of local air pollution.

Like air pollution, noise pollution is also highly localised and area dependent and its measurement is difficult and expensive. In the context of rail-based transportation, noise nuisance is generally associated with network construction, depot activity and the noise of passing trains affecting adjacent properties. Where modal switch is significant, there may be significant implications for the change in the level of noise from road traffic, especially in the context of the number of cars in locally affected areas. Regarding noise pollution, there are two key issues of interest: first, compared to air pollution, noise pollution tends to have less severe effects on physical human health [133]. The second point of concern is the evidence showing that people tend to find noise from rail transportation less annoying than other modes of transportation [134]. The evaluation for noise pollution can be based on the number of people exposed to certain noise levels for the baseline and *with rail* scenarios. Some examples of noise exposure modelling can be seen in HEATCO and UNITE projects which utilised either WTP or Hedonic Price studies for the valuation of noise exposure (see also EC, [111]).

Table 7. Examples of damage costs of main pollutants from transport, € per tonne, 2010.

Country	NO_x	NMVOC	SO_2	$PM_{2.5}$		
				Rural	Suburban	Urban
France	13,052	1695	12,312	33,303	64,555	211,795
Germany	17,039	1858	14,516	48,583	73,221	222,461
Ireland	5688	1398	6959	16,512	47,420	194,660
Italy	10,824	1242	9875	24,562	50,121	197,361
Netherlands	11,574	2755	16,738	29,456	48,352	195,592
Poland	13,434	1678	14,435	47,491	74,215	221,455

Source: (see) EC [111] for all EU country values.

4.6. Public service provision costs

Public service provision costs are widely used as an indicator to demonstrate how costs of public service provision tend to increase with dispersion of urban activities. The literature has reached a consensus on this issue as there is growing body of research indicating that dispersed population expansion increases the costs of local public service provision [70, 135]. Research on increased density developments indicates such development tends to reduce per capita costs of providing public infrastructure and services [136]. Concerning land development impacts of rapid rail investments, building such infrastructure developments is generally accepted as preferable to an alternative to urban development that supports car- based travel [137]. The reason is that rapid rail developments can provide high- quality services in terms of reliability, speed, safety and reduced travel time, and can support higher density developments along the catchment area as required for the efficiency of rail service provision [136].

In order to represent land development impacts of rapid rail provisions, public service provision costs can be utilised as an indicator for the cost-benefit evaluation of each scenario indicating a negative cost impact of the automobile-oriented development in the baseline scenario compared to an urban form resulting from proposed rail investment and integrated land use/transport planning policies. Though this is not a commonly used indicator in the EU-wide transport project appraisals, the inclusion of this indicator in the CBA model is vital to represent possible benefits accrued to urban form resulting from rail transport investments and associated planning policies. Public service cost estimations are case specific and can be identified as the costs of road construction, housing and community facilities development, education, fire and police protection, water and electricity distribution, sewerage, and social and recreational facilities [136]. Unit public service costs can be computed where data are available on each of these different cost items.

Europe-wide research on a common assessment methodology concerning changes in public service provision costs stemming from a rail-based urban development and planning is limited. EC's project series such as ExternE [130] deal with a wide range of internal and external impacts of transport projects and policy changes in EU countries. However, research on land use impacts of major transport investments has not been fully covered. An EC [138] Report on Thematic Strategy on the Urban Environment Impact Assessment outlines integrated strategies for member states to the sustainable management of urban environment. In this report, the consideration of urban environment indicators relating to sustainable urban development and transportation systems were addressed for an integrated assessment and management of the urban environment across EU member states. There are European projects such as PROPOLIS and SCATTER focusing on the linkages between urban form and urban transport systems and the impacts of urban trans- port and land use policies. The outcomes of these projects contribute to the existing research on the issues of land use impacts of transport provisions in the European cities. However, case studies in these projects cover a limited number of urban areas and, therefore, more comprehensive analysis with wider European area coverage is needed. Therefore, this can be specified as a priority topic for future research con- siderations to develop a common assessment methodology for the subject indicator.

5. Cost-benefit evaluation

Given the methodological framework specified above for the assessment of key impacts and indicators, costs and benefits of rail investments can be calculated and assessed between two scenarios of baseline and *with rail* cases. The development of these scenarios is based on some European seminal sources highlighting scenario analysis as a tool to be used for the policy analysis in the EU [138]. As previously stated, a baseline scenario assumes that urban area would continue to grow with the present trends with sufficient maintenance and renewal investments to maintain existing infrastructure. A *with rail* scenario, on the other hand, consists of at least one new rail transport investment connected to the existing infrastructure in the area.

A general CBA approach implies that all costs and benefits are reduced to their present value and discounted at a standard rate over the pre-specified evaluation period through the formula given below:

$$ENPV = \sum_{t=0}^{n} a_t S_t = \frac{(b_0 - c_0)}{(1+r)^0} + \frac{(b_1 - c_1)}{(1+r)^1} + \ldots + \frac{(b_n - c_n)}{(1+r)^n} \quad (1)$$

where S_t is the balance of cash flow funds comprising flow of benefits b_t and flow of costs C_t; a_t is the discount factor; r is the discount rate and n is the evaluation period [8]. This is also used to produce a benefit-cost ratio (BCR) and internal rate of return (IRR). The former is the ratio of the discounted aggregate net benefits (i.e., benefits minus costs) to the discounted investment costs and the latter is the rate of discount equating discounted net benefits to discounted investment costs.

There are differences among decision criteria used by different EU countries. Odgaard et al. [47] noted that all EU states, except Finland and Sweden, apply more than one decision rule for the CBA evaluation of a project. Among them, ENPV and BCR are the most widely used, which is followed by the IRR.

The social discount rates and project appraisal periods vary among countries reflecting the local variations in opportunity costs of capital, project risks and life- times of rapid rail investments. The UNITE project suggests the use of a European social discount rate of 3% while EC HEATCO suggests a rate of 5% (HEATCO D5, [105]). This implies the use of a range of discount rates between 3 and 5% in CBA. The use of a specific discount rate and an evaluation period depending on the characteristics of the project and the national assessment procedure is recom- mended for each specific country. In the absence of information regarding the evaluation period in national appraisal guidelines, the evaluation period of 40 years is suggested (as in HEATCO) as a default evaluation period (i.e., planning and construction period plus 40 years of operational period).

6. Conclusion

In conclusion, this research has reviewed literature and summarised the key methodological approaches in relation to cost-benefit analysis of transport investment projects and programmes used in the testing of various scenarios relating to urban form. The use of at least two different land development scenarios (such as *business- as-usual* and *with rail* cases) is required to allow the CBA process to be used in discussions of alternative development and investment decisions linked to urban form and development issues. Transformation from compact to more dispersed structures has significant implications on the urban environment and is generally associated with high social, economic and environmental costs. To address some of these problems, planning theory and practice have increased their focus on issues of sustainable urban development and urban growth management. In this context, the CBA approach is intended to allow for the development of an improved quantitative evidence base for decisions on infrastructure spending considering the potential costs and benefits of such an investment decision in comparison with alternative options and scenarios.

Regarding the selected inputs, we conclude that the issue of underestimation is a recurring feature of public transport projects, and based on the findings of literature, different averages of cost escalations can be added to national cost estimates based upon European average cost escalations. An issue highlighted in this paper is the potential transferability of the input data on vehicle operation costs between countries due to the existence of similarities in the operating cost relationships for road vehicles. Regarding national values for work and non-work travel time savings, we conclude that these can be identified through meta-analysis models in order to determine travel time values for the country of interest. Regarding values of traffic safety, the concept of inter-temporal elasticity to GDP has been introduced to estimate the future values. However, this issue seems to be less relevant considering the present economic climate of recession. An important conclusion relates to the area price impacts and wider economic benefits which are routinely not accounted for due to displacement issues despite their obvious importance. Considering environmental impacts, the use of carbon charges is suggested for the valuation, and environmental indicators can be estimated based on national labour cost data and inter-modal basis. Public service provision costs are more difficult to establish; however, it can be noted that evidence of utility provision costs provides useful data for the valuation of this input. **Table 8** summarises the main impacts and relative evaluation methods to be considered for the economic appraisal of transport infrastructure projects.

This chapter has reviewed and outlined key base indicators which can be utilised in a CBA methodology with information on data sources used in EU countries.

Clearly, the methodology is adaptable in the sense that in specific states, other parameters that are crucial to any cost-benefit analyses can be added or removed. In this sense, this review suggests that the CBA methodology can produce a common set of base indicators which is sufficiently flexible to be adapted to country-specific contexts across the EU.

Table 8. Valuation methods of key impacts/indicators of public transport provisions.

Impacts/indicators	**Valuation methods**
Travel time savings	• Stated preferences • Revealed preferences (multi-purpose household/business surveys) • Cost saving approach
Vehicle operation cost savings	• Market value
Operating costs of carriers	• Market value
Accident cost savings	• Stated preferences • Revealed preferences (hedonic wage method) • Human capital approach
Public service provision cost savings	• Market value
Costs savings: noise emissions	• WTP/WTA compensation • Hedonic price method
Cost savings: local air pollution	• Shadow price of air pollutants
Cost savings: GHG emissions	• Shadow price of GHG emissions
Source: EC [9].	

References

1. Cervero R. Transport Infrastructure and the Environment: Sustainable Mobility and Urbanism. Working Paper 2013-03. US: University of California, Institute of Urban and Regional Development; 2013

2. Koglin T, Rye T. The marginalisation of bycling in modernist urban transport planning. Journal of Transport and Health. 2014;1(4):214-222

3. Kaldor N. Welfare propositions of economics and interpersonal comparisons of utility. Economic Journal. 1939;49:549-552

4. Annema JA, Koopmans C, Wee Van B. Evaluating transport infrastructure investments: The Dutch experience with a standardised approach. Transport Reviews. 2007;27(2):125-150

5. Mackie P, Worsley T. International Comparisons of Transport Appraisal Practice: Overview Report. UK: University of Leeds, Institute for Transport Studies; 2013
6. Jones H, Moura F, Domingos T. Transport infrastructure project evaluation using cost-benefit analysis. Procedia-Social and Behavioural Sciences. 2014;111:400-409
7. European Commission (EC). Cost- Benefit and Multi-Criteria Analysis for New Road Construction: Transport Research EURET Concerted Action 1.1. Brussels: EC; 1996
8. European Commission (EC). Directorate General Regional Policy. Guide to Cost Benefit Analysis of Investment Projects: Final Report 2008. Brussels: EC; 2008
9. European Commission (EC). Directorate General Regional Policy. Guide to Cost Benefit Analysis of Investment Projects: Economic Appraisal Tool for Cohesion Policy 2014-2020. Brussels: EC; 2014
10. Bristow AL, Nellthorp J. Transport project appraisal in the European Union. Transport Policy. 2000;7(1):51-60
11. Bueno PC, Vassallo JM, Cheung K. Sustainability assessment of transport infrastructure projects: A review of existing tools and methods. Transport Reviews. 2015;35(5):622-649
12. Delbosc A, Currie G. The spatial context of transport disadvantage, social exclusion and well-being. Journal of Transport Geography. 2011;19:1130-1137
13. Bhattacharjee S, Goetz AR. Impact of light rail on traffic congestion in Denver. Journal of Transport Geography. 2012;**22**:262-270
14. Czaika M, de Haas H. The globalization of migration: Has the world become more migratory? International Migration Review. 2014;48(2):283-323
15. Cairncross F. The Death of Distance: How the Communications Revolution Is Changing Our Lives. Boston: Harvard Business School Press; 2001
16. Scott A. Global City-Regions: Trends, Theory, Policy. Oxford: Oxford University Press; 2001
17. Anas A, Arnott R, Small KA. Urban spatial structure. Journal of Economic Literature. 1998;36:1426-1464
18. Breheny M. Compact cities and transport energy consumption. Transactions of the Institute of British Geographers. 1995;20(1):81-101
19. Hillman M. In favour of the compact city. In: Jenks M, Burton E, Williams K, editors. The Compact City: A Sustainable Urban Form? Oxford: Spon Press; 1996
20. Williams K. Spatial planning, urban form and sustainable transport: An introduction. In: Williams K, editor. Spatial Planning, Urban Form and Sustainable Transport. Hampshire, Ashgate; 2005
21. Williams K, Jenks M, Burton E. Achieving Sustainable Urban Form. London: E and F N Spon; 2000
22. Bertaud A. The Spatial Organization of Cities: Deliberate Outcome or Unforeseen Consequence? Working Paper 2004-01. Berkeley, Institute of Urban and Regional Development, University of California; 2004

23. Westerink J, Haase D, Bauer A, Ravetz J, Jarrige F, Aalbers CB. Dealing with sustainability trade-offs of the compact city in peri-urban planning across European City regions. European Planning Studies. 2013;21(4):473-497

24. Gordon P, Richardson HW, Jun MI. The commuting paradox: Evidence from the top twenty. Journal of the American Planning Association. 1991;(4):416-420

25. Gordon P, Kumar A, Richardson HW. Congestion, changing metropolitan structure, and city size in the United States. International Regional Science Review. 1989;12(1):45-56

26. Aguilera A. Growth in commuting distances in French polycentric metropolitan areas: Paris, Lyon and Marseille. Urban Studies. 2005;42(9):1537-1547

27. Ma K-R, Banister D. Urban spatial change and excess commuting. Environment and Planning A. 2007;39:630-646

28. Van der Laan L. Changing urban systems: An empirical analysis at two spatial levels. Regional Studies. 1998;32(3):235-247

29. Donaldson R. Mass rapid rail development in South Africa's Metropolitan Core: Towards a new urban form? Land Use Policy. 2005;23:344-352

30. Perl A. Assessing the recent reformulation of United States passenger rail policy. Journal of Transport Geography. 2012;22:271-281

31. Ustaoglu E, Williams B, Murphy E. Integrating CBA and land-use development scenarios: Evaluation of planned rail investments in the Greater Dublin Area, Ireland. Case Studies on Transport Policy. 2016;4:104-121

32. Ustaoglu E, Williams B, Petrov OL, Shahumyan H, van Delden H. Developing and assessing alternative land-use scenarios from the MOLAND model: A scenario-based impact analysis approach for the evaluation of rapid rail provisions and urban development in the Greater Dublin Region. Sustainability. 2018;10:61

33. Debrizon G, Pels E, Rietveld P. The effects of railway investments in a polycentric city: A comparison of competitive and segmented land markets. Environment and Planning A. 2007;39:2048-2067

34. Willigers J, Floor H, Wee Van B. Accessibility indicators for location choices of offices: An application to the intraregional distributive effects of high-speed rail in the Netherlands. Environment and Planning A. 2007;39:2086-2098

35. Geurs KT, Wee Van B, Rietveld P. Accessibility appraisal of integrated land-use-transport strategies: Methodology and case study for the Netherlands Randstad Area. Environment and Planning B. 2006;33:639-660

36. Pearce D, Atkinson G, Mourato S. Cost-Benefit Analysis and the Environment: Recent Developments. Paris: OECD Publishing; 2006

37. Litman T. What's It Worth? Economic Evaluation for Transportation Decision-Making [Internet]. . Victoria Transport Policy Institute. 2006. Available from: http://www.vtpi.org [Accessed: December 24, 2018]

38. Sinha CK, Labi S. Transportation Decision Making: Principles of Project Evaluation and Programming. New Jersey: John Wiley & Sons; 2011

39. OECD. Cost Benefit Analysis and the Environment: Recent Developments. Paris: OECD; 2006

40. Brent RJ. Applied Cost-Benefit Analysis. UK: Edward Elgar Publishing; 2008

41. World Bank. Where to Use Cost Effectiveness Techniques Rather than Cost-Benefit Analysis. Transport Notes No. TRN-9. Washington, DC: Transport Economics, Policy and Poverty Thematic Group. Washington, DC: World Bank; 2005

42. Flintsch GW. Life-cycle assessment as a tool for sustainable transportation Infrastructure management. In: Environmentally Conscious Transportation. Hoboken, NJ, USA: John Wiley & Sons Inc.; 2008

43. Vreeker R, Nijkamp P. Multicriteria evaluation of transport policies. In: Button KJ, Hensher DA, editors. Handbook of Transport Strategy, Policy and Institutions. The Netherlands: Elsevier; 2005

44. Banai R. Land resource sustainability for urban development: Spatial decision support system prototype. Environmental Management. 2005;36(2):282-296

45. Hensher DA, Button KJ, Haynes KE, Stopher PR, editors. Handbook of Transport Geography and Spatial Systems: Handbooks in Transport. Vol. 5. Amsterdam: Elsevier; 2004

46. Grant-Muller SM, Mackie P, Nellthorp J, Pearman A. Economic appraisal of European transport projects: The state-of-the-art revisited. Transport Reviews. 2001;21(2):237-261

47. Odgaard T, Kelly C, Laird J. Current Practice in Project Appraisal in Europe: Analysis of Country Reports. HEATCO D:1 [Internet]. 2005. Available from: http://heatco.ier.uni-stuttgart.de/ [Accessed: October 21, 2018]

48. OECD. Impact of Transport Infrastructure Investment on Regional Development. Paris: OECD; 2002

49. Banister D, Thurstain-Goodwin M. Quantification of the non-transport benefits resulting from rail investment. Journal of Transport Geography. 2011;19(2):212-223

50. Nijkamp P, Rietveld P, Voogd H. Multicriteria Evaluation in Physical Planning. Amsterdam: North Holland; 1991

51. Beuthe M, Eeckhoudt L, Scannella G. A practical multicriteria methodology for assessing risky public investments. Socio-Economic Planning Sciences. 2000;34(2):121-139

52. Litman T. Evaluating Transportation Land Use Impacts [Internet].. Victoria Transport Policy Institute. 2019. Available from: http://www.vtpi.org/ landuse.pdf [Accessed: April 12, 2019]

53. Banister D, Berechman J. Transport Investment and Economic Development. UK: UCL Press; 2000

54. EEA. TERM 2006-Transport and Environment: On the Way to a New Common Transport Policy-Indicators Tracking Transport and Environment in the European Union. EEA Report no.1, Copenhagen; European Environmental Agency; 2007

55. Gibbons S, Lyytikainen T, Overman H, Sanchis-Guarner R. New Road Infrastructure: The Effects on Firms. SERC Paper No. 117. London: London School of Economics; 2012

56. Graham D. Wider Economic Benefits of Transport Improvements: Link between City Size and Productivity. London: DfT; 2005

57. Venables A. Evaluating Urban Transport Improvements: Cost Benefit Analysis in the Presence of Agglomeration and Income Taxation, CEP Discussion Paper No651. London: London School of Economics and Political Science; 2004

58. Fujita M, Thisse J. Economics of Agglomeration. Cambridge: Cambridge University Press; 2002

59. Crafts N. Transport infrastructure investment: Implications for growth and productivity. Oxford Review of Economic Policy. 2009;25(3):327-343

60. Department for Transport (DfT) UK. Transport, Wider Economic Benefits, and Impacts on GDP. London: HMSO; 2005

61. Nash C, Preston J. Appraisal of rail investment projects: Recent British experience. Transport Reviews. 1991;11(4):295-309

62. Duncan M. The impact of transit- oriented development on housing prices in San Diego, CA. Urban Studies. 2011;48(1):101-127

63. Knight RL, Trygg LL. Evidence of land use impacts of rapid transit systems. Transportation. 1977;6:231-247

64. Commission of the European Communities. Green Paper on Towards a New Culture for Urban Mobility. Brussels: EC; 2007

65. Chapman L. Transport and climate change: A review. Journal of Transport Geography. 2007;15(5):354-367

66. Proost S, De Borger B, Koskenoja P. Public finance aspects of transport charging and investments. In: De Palma A, Lindsey R, Proost S, editors. Research in Transportation Economics. UK: Elsevier; 2007

67. Dahlby B. The Marginal Cost of Public Funds. Cambridge, MA: The MIT Press; 2008

68. Magnussen OM, Olsson NO. Comparative analysis of cost estimates of major public investment projects. International Journal of Project Management. 2006;24(4):281-288

69. Cervero R, Kang CD. Bus rapid transit impacts on land uses and land values in Seoul, Korea. Transport Policy. 2011;18(1):102-116

70. Carruthers IJ, Ulfarsson GF. Does `smart growth` matter to public finance? Urban Studies. 2008;45(9):1791-1823

71. Buehler R, Pucher J. Sustainable transport in Freiburg: Lessons from Germany's environmental capital. International Journal of Sustainable Transportation. 2011;5:43-70

72. Gospodini A. Urban development, redevelopment and regeneration encouraged by transport infrastructure projects: The case study of 12 European cities. European Planning Studies. 2005;13(7):1083-1111

73. Falkmer T, Renner L, Anund A. Estimated societal costs of a hierarchical measures approach to enhanced school transportation safety at bus stops on roads with high speed limits. European Transport Research Review. 2010;2(4):201-207

74. Janic M. Modelling the full costs of an intermodal and road freight transport network. Transportation Research D. 2007;12(1):33-44

75. Calvo F, Oña De J. Are rail charges connected to costs? Journal of Transport Geography. 2012;22:28-33

76. Börjesson M, Eliasson J, Franklin JP. Valuations of travel time variability in scheduling versus mean-variance models. Transportation Research B. 2012;46(7):855-873

77. Nkurunziza A, Zuidgeest M, Brussel M, Van den Bosch F. Spatial variation of transit service quality preferences in Dar-es-Salaam. Journal of Transport Geography. 2012;24:12-21

78. Chang JS. Estimation of option and non-use values for intercity passenger rail services. Journal of Transport Geography. 2010;18:259-265

79. Tiwari G, Jain D. Accessibility and safety indicators for all road users: Case study Delhi BRT. Journal of Transport Geography. 2012;22:87-95

80. Hou Q , Li S-M. Transport infrastructure development and changing spatial accessibility in the Greater Pearl River Delta, China, 1990- 2020. Journal of Transport Geography. 2012;19(6):1350-1360

81. Olvera LD, Plat D, Pochet P. Household transport expenditure in sub-Saharan African cities: Measurement and analysis. Journal of Transport Geography. 2008;16:1-13

82. Chen C-L, Hall P. The impacts of high-speed trains on British economic geography: A study of the UK's InterCity and its effects. Journal of Transport Geography. 2011;19(4):689-704

83. Hensher DA, Truong TP, Mulley C, Ellison R. Assessing the wider economy impacts of transport infrastructure investment with an illustrative application to the north-west rail link project in Sydney, Australia. Journal of Transport Geography. 2012;24:292-305

84. Geurs KT, Zondag B, de Jong G, de Bok M. Accessibility appraisal of land use/transport policy strategies: More than just adding up travel-time savings. Transportation Research D. 2010;15(7):382-393

85. Proost S, Sen A. Urban transport pricing reform with two levels of government: A case study of Brussels. Transport Policy. 2006;13(2):127-139

86. De Palma A, Proost S, van der Loo S. Assessing transport investments—Towards a multi-purpose tool. Transportation Research B. 2010;44(7):834-849

87. Loon van R, Rietveld P, Brons M. Travel-time reliability impacts on railway passenger demand: A revealed preference analysis. Journal of Transport Geography. 2011;19:917-925

88. Hadas Y, Ranjitkar P. Modeling public-transit connectivity with spatial quality-of-transfer measurements. Journal of Transport Geography. 2012;22:137-147

89. Hensher DA. Sustainable public transport systems: Moving towards a value for money and network- based approach and away from blind commitment. Transport Policy. 2007;14(1):98-102

90. Ong HC, Mahlia TM, Hasjuki H. A review on emissions and mitigation strategies for road transport in Malaysia. Renewable and Sustainable energy Reviews. 2011;15(8):3516-3522

91. Suau-Sanchez P, Pallares-Barbera M, Paul W. Incorporating annoyance in airport environmental policy: Noise, societal response and community participation. Journal of Transport Geography. 2011;19(2):275-284

92. Lefevre B. Long-term energy consumptions of urban transportation: A prospective simulation of 'transport- land uses' policies in Bangalore. Energy Policy. 2009;37(3):940-953

93. Li M-H, Barrett M, Rammohan P, Olivera F, Landphair H. Documenting storm water quality on Texas highways and adjacent vegetated roadsides. Journal of Environmental Engineering. 2008;134(1):48-59

94. Tempelman E. Multi-parametric study of the effect of materials substitution on life cycle energy use and waste generation of passenger car structures. Transportation Research D. 2011;16:479-485

95. Antonson H, Gustafsson M, Angelstam P. Cultural heritage connectivity: A tool of EIA in transportation infrastructure planning. Transportation Research D. 2010;15:463-472

96. Janic M. (2003) multicriteria evaluation of high-speed rail, transrapid Maglev and air passenger transport in Europe. Transportation Planning and Technology. 2003;26(6):491-512

97. Litman T. Well Measured: Developing Indicators for Comprehensive and Sustainable Transport Planning. [Internet] Victoria Transport Policy Institute. 2019. Available from: http://www.vtpi.org/ wellmeas.pdf [Accessed: April 12, 2019]

98. Litman T, Burwell D. Issues in sustainable transportation. International Journal of Global Environmental Issues. 2006;**6**(4):331-347

99. Marsden G, Kimble M, Nellthorp J, Kelly C. Sustainability assessment: The definition deficit. International Journal of Sustainable Transportation. 2010;4:189-211

100. UN. Sustainable Transport Evaluation. UN Department of Economic and Social Affairs. CSD19/2011/BP10. Eschborn, Germany: UN; 2011

101. Flyvbjerg B, Holm MS, Buhl S. Underestimating costs in public works projects: Error or lie? Journal of the American Planning Association. 2002;68(3):279-295

102. Flyvbjerg B. COWI. Procedures for Dealing with Optimism Bias in Transport Planning: Guidance Document. London: UK Department for Transport; 2004

103. Love PED, Zhou J, Irani Z, Edwards DJ, Sing CP. Off the rails: Cost performance of rail infrastructure projects. Transportation Research A. 2017;99:14-29

104. Love PED, Ahiaga-Dagbui D, Welde M, Odeck J. Cost performance light transit rail: Enablers of future- proofing. Transportation Research A. 2017;100:27-39

105. HEATCO. Funded by 6th Framework Programme IER, Germany. Brussels: European Commission EC-DG TREN; 2004-2006

106. de Blaeij A, Koetse M, Tseng YY, Rietveld P, Verhoef E. Valuation of Safety, Time, Air Pollution, Climate Change and Noise; Methods and Estimates for Various Countries. Report for ROSEBUD. Amsterdam: Department of Spatial Economics, Vrije Universiteit; 2004

107. Wijnen W, Weijermars W, Vanden Berghe W, Schoeters A et al. Crash Cost Estimates for European Countries. Deliverable 3.2 of the H2020 Project SafetyCube. UK: Loughborough University Institutional Repository; 2017

108. Viscusi WK, Aldy JE. The value of a statistical life: A critical review of market estimates throughout the world. Journal of Risk and Uncertainty. 2003;27(1):5-76

109. Treich N. The value of a statistical life under ambiguity aversion. Journal of Environmental Economics and Management. 2010;59(1):15-26

110. Woods B, Revill P, Sculpher M, Claxton K. Country-level cost- effectiveness thresholds: Initial estimates and the need for further research. Value in Health. 2016;19(8):929-935

111. European Commission (EC). DG Mobility and Transport. Update of the Handbook on External Cost of Transport. Final Report. Ricardo AEA/ ED57769. Brussels: EC; 2014

112. Bowland BJ, Beghin JC. Robust estimates of value of a statistical life for developing economies. Journal of Policy Modelling. 2001;23:385-396

113. Litman T. Evaluating Public Transport Benefits and Costs [Internet].. Victoria Transport Policy Institute. 2019. Available from: http://www.vtpi.org/ tranben.pdf [Accessed: April 12, 2019]

114. Archondo-Callao R. Applying the HDM-4 Model to Strategic Planning of Road Works. World Bank Transport Papers TP-20. Washington, DC: World Bank; 2008

115. Talvitie A. Evaluation of road projects and programs in developing countries. Transport Policy. 2000;7:61-72

116. Cutura B, Mladenovic G, Mazic B, Lovric I. Application of the HDM-4 model on local road network: Case study of the Herzagovina-Neretva Canton in Bosnia and Herzegovina. Transportation Research Procedia. 2016;14:3021-3030

117. Perrotta F, Parry T, Neves LC, Buckland T, Benbow E, Mesgarpour M. Verification of the HDM-4 fuel consumption model using a big data approach: A UK case study. Transportation Research D. 2019;67:109-118

118. World Bank. Notes on the Economic Evaluation of Transport Projects. Transport Economics, Policy and Poverty Thematic Group, TRN-14. Washing[1], DC: World Bank; 2005

119. Anderson P. Understanding the Costs of Commercial Railways. Transport Infrastructure Note RW-10. Washington, DC: World Bank; 1995

120. Hensher DA. Value of Business Travel Time. London: Permagon Press; 1977

121. Fowkes AS, Marks P, Nash CA. Business travel. In: Fowkes AS, Nash CA, editors. Analysing Demand for Rail Transport. Avebury: Aldershot; 1991

122. Mackie PJ, Jara-Diaz S, Fowkes AS. The value of travel time savings in evaluation. Transportation Research E. 2001;37:91-106

123. Wardman M. The value of travel time: A review of British evidence. Journal of Transport Economics and Policy. 1998;32(3):285-316

124. Wardman M, Chintakayala P, de Jong G, Ferrer D. European Wide Meta- Analysis of Values of Travel Time. UK: University of Leeds; 2012

125. Tol RSJ, Lyons S. Incorporating GHG Emission Costs in the Economic Appraisal of Projects Supported by State Development Agencies. Working Paper 247. ESRI: Ireland; 2008

126. Kuik O, Brander L, RSJ T. Marginal Abatement Costs of Carbon-Dioxide Emissions: A Meta Analysis. Working Paper 248. ESRI: Ireland; 2008

127. Havranek T, Irsova Z, Janda K, Zilberman D. Selective reporting and the social cost of carbon. Energy Economics. 2015;51:394-406

128. Tol RSJ. The marginal damage costs of carbon dioxide emissions: An assessment of the uncertainties. Energy Policy. 2005;33(16):2064-2074

129. Watkiss P, Anthoff D, Downing T, Hepburn C, Hope CHR, Hunt A, et al. The social cost of carbon (SCC) review. Final Report November 2005. 2005

130. European Commission (EC). ExternE-Externalities of Energy– Methodology 2005 Update. EUR21951 EN. Luxembourg: Office for Official Publications of the European Communities; 2005

131. Preiss P, Klotz V. Description of Updated and Extended Draft Tools for the Detailed Site-Dependant Assessment of External Costs. Technical Paper No. 7.4. RS1b of NEEDS Project. 2007

132. Krzyzanowski M, Kuna-Dibbert B, Schneider J. Health Effects of Transport- Related Air Pollution. Copenhagen: WHO; 2005

133. WHO. Experts Consultation on Methods of Quantifying Burden of Disease Related to Environmental Noise. Stuttgart: State Health Agency of Baden-Württemberg; 2005

134. King EA, Murphy E, Rice HJ. Implementing the EU environmental noise directive: Lessons from the first phase of strategic case mapping and action planning in Ireland. Journal of Environmental Management. 2011;92(3):756-764

135. Buettner T, Schwager R, Stegarescu D. Agglomeration, population size and the cost of providing public services: An empirical analysis for German states. Public Finance and Management. 2004;4(4):496-520

136. Litman T. Understanding Smart Growth Savings. [Internet]. Victoria Transport Policy Institute. 2011. Available from: http://www.vtpi.org/ sg_save.pdf [Accessed: October 12, 2018]

137. Vuchie VR. Recognizing the value of rail transit. T R News. 1991;156:13-19

138. European Commission (EC). Communication from the Commission to the Council and the European Parliament on Thematic Strategy on the Urban Environment Impact Assessment. Brussels: EC; 2006

Index